Smart Home Systems

Smart Home Systems

Edited by
Mahmoud A. Al-Qutayri

Smart Home Systems
Edited by Mahmoud A. Al-Qutayri

CBS, Edition **2017**

Published by InTech
Janeza Trdine 9, 51000 Rijeka, Croatia

Publishing Process Manager Sandra Bakic

Technical Editor InTech DTP team

Cover Designer InTech Design Team

Additional hard copies can be obtained from orders@intechopen.com

Smart Home Systems, Edited by Mahmoud A. Al-Qutayri
p. cm.
ISBN 978-953-307-050-6

Preface

The tremendous developments that continue to evolve in the field of information communication technology (ICT) are affecting nearly all aspects of human lives. Mobile phones currently support many applications and deliver services that go well beyond basic voice communication. Wireless sensor networks with ubiquitous embedded processors can sense various parameters and generate the appropriate responses. These are some examples of systems that are being driven by advancements in ICT. Another important ICT driven field is the smart home. This field has been the subject of concerted research effort for some time and is beginning to develop practical products for a market that is expected to grow substantially in the relatively near future.

Smart homes are intelligent environments that interact dynamically and respond readily in an adaptive manner to the needs of the occupants and changes in the ambient conditions. The realization of systems that support the smart homes concept requires the integration of technologies from different fields. Among the challenges that the designers face is to make all the components of the system interact in a seamless, reliable and secure manner. Another major challenge is to design the smart home in a way that takes into account the way humans live and interact. This later aspect requires input from the humanities and social sciences fields. The need for input from diverse fields of knowledge reflects the multidisciplinary nature of the research and development effort required to realize smart homes that are acceptable to the general public.

The applications that can be supported by a smart home are very wide and their degree of sophistication depends on the underlying technology used. Some of the application areas include monitoring and control of appliances, security, telemedicine, entertainment, location based services, care for children and the elderly... etc.

This book consists of eleven chapters that cover various aspects of smart home systems. Chapter 1: *Smart Home Systems* introduces pervasive computing in the context of smart homes and then presents a vision of the home environment, and how devices, networks and applications should be organized. It also describes the use of a home gateway to support the home applications execution. Chapter 2: *Integrated Wireless Technologies for Smart Homes Applications* describes some of the key wireless communication, smart sensors, and context-aware computing technologies used to design smart homes. It then presents various smart homes applications and describes the design and implementation of an integrated secure application that enables monitoring and control of multiple devices with multiple settings at a remote home using a mobile phone. Chapter 3: *Selected Home Automation and Home Security Realizations: An Improved Architecture* presents a study on the control of some appliances, timed setting of switching the appliances and personal digital home assistant software that brings the attention of the resident about the tasks of the day to be performed. The study includes various forms of home access security. It also describes a dual energy extraction unit generating electricity from the renewable energy resources that can be incorporated in the overall system.

Chapter 4: *Applying Agent-based Theory to Adaptive Architectural Environments – Example of Smart Skins* presents a study that employs intelligent agent theory to investigate an adaptive architectural environment, takes smart skins as a research example, and proposes the use of neuro-fuzzy to establish judgment and reaction control conditions. The study derives adaptive environmental research categories, establishes adaptive building environment hypotheses based on intelligent agent theory, summarizes and analyzes the numerous feasible computational mechanisms and selection conditions, and employs a prototype smart skin structure to design experiments, perform testing and assessment, and analyzes the relationship between users, the environment, and the smart skin. Chapter 5: *An AmI-enabled OSGi Platform Based on Socio-semantic Technologies* introduces an approach for context-aware personalized smart homes based on the open service gateway initiative (OSGi) platform, the semantic-web philosophy and the collaborative tagging trends. The combination of these fields provides a semantical conceptualization of services at home. The architecture described in the study also supports context-awareness personalization by using a dynamic and non-previously agreed structure for modelling context. This socio-semantic approach to the problem takes into account the heterogeneous nature of the devices which provide contextual information so that defining a previously agreed constrained vocabulary for context is unrealistic.

Chapter 6: *Aging in Place: Self-Care in Smart Home Environments* discusses the merits of people aging in the place where they are used to live and the burden that this places on caring for them. It then describes the components and functionalities of smart homes that can be used to support aging in place. Chapter 7: *Telemonitoring of the Elderly at Home: Real-time Pervasive Follow-up of Daily Routine, Automatic Detection of Outliers and Drifts* discusses the ability to obtain reliable pervasive information at home from a network of localizing sensors allowing following the different activity-station at which a dependent person, such as an elderly, can be detected. The main idea is to watch the person at home in order to classify its activities of daily living, detect early its abnormal states or behaviour like perseveration, diagnosis and prolong its autonomy. The study proposes two mathematical approaches for location modelling. The data collection, modelling, numerical analysis, and decision making are used within the context of the smart home for monitoring the elderly persons. Chapter 8: *Smart Home with Healthcare Technologies for Community-Dwelling Older Adults* discusses societal needs for the technology and the current smart homes status. It then presents a review of older adults' perceptions of smart homes, healthcare technologies and their effectiveness, and caregivers' perspectives of the technology. It also defines the role of smart homes with health care technologies in the model toward optimal management for independent through technological adoption. The model rests on behavioural medicine to promote healthy behaviours in community-dwelling older adults.

Chapter 9: *Memory Management in Smart Home Gateway* discusses the memory management in gateways and prioritizing the memory use to maximize the number of services running simultaneously in the home gateway. It highlights the differences between memory management for software bundles executed in home gateways and traditional memory management techniques. It then proposes new algorithms for efficient management of service bundles and presents an evaluation of the results achieved. Chapter 10: *Virtual Place Framework for User-centered Smart Home Applications* proposes a series of smart home platforms which enables home users to experience a smart virtual place through the Internet offered by the project architect. The proposed interactive virtual place is different from conventional 3D space in that the created virtual place embodies spatial context-aware information including spatial relationship among entities, activities and users.

I would like to thank and all those who contributed to making this book possible. They include the authors who submitted their book chapters and the editorial and production staff of IN-TECH Education and Publishing. I hope the readers will find this book on smart homes informative and enjoyable to read.

February 2010

Mahmoud A. Al-Qutayri
Khalifa University of Sci., Tech. and Research
UAE

Contents

1

Smart Home Systems

P. LALANDA, J. BOURCIER, J. BARDIN & S. CHOLLET
Grenoble University
FRANCE

1. Introduction

The pervasive computing area has recently gained major importance from both industry and academia and is changing the way we interact with our environment [1]. This computing domain emphasizes the use of small, intelligent and communicating daily life objects to interact with the computing infrastructure. These devices tend to blend in their environment. This is especially true in our homes where new electronic devices such as photo frames aim to be as decorative as powerful. Devices are then not always perceivable by human beings. These new equipments have the ability to communicate with each other, to configure or repair themselves, and perform context-based cognitive and physical actions. The vision of coordinated or cooperating devices teaming up transparently to provide human beings with services of all sorts is actually getting closer and closer.

However, the main part of research efforts has focused so far on providing hardware that can actually enable such interactions. Consequently, plenty of devices providing this kind of features are already commercialized, whereas very few interesting applications take advantages of this new infrastructure.

Indeed, the complexity of building software that can actually benefit from this underlying hardware is often underestimated. Usual software engineering techniques and tools are not suitable. Several software engineering challenges remain to be solved before fulfilling the vision of a true pervasive world. Notably the high degree of dynamism, distribution, heterogeneity and autonomy of the devices involved raises important problems. Major security and privacy concerns have also to be considered while building such systems. Indeed, the Home Network emphasizes the envisioned environment openness to networked entities. It is open to dynamic connections: devices may enter and leave the network spontaneously, providing context-dependent features (e.g. according to user's activity). It is also open to heterogeneous devices: protocols and device types differ according to application domains and service providers. Moreover, devices are spread over the home space, which is not clearly delimited for wireless communicating devices.

In this chapter we describe our work dealing with the provision of a natural execution environment simplifying the construction of pervasive applications. More specifically, we describe our vision of smart home environments, and propose an infrastructure to support the execution of home applications providing end user services using different features provided on the home network.

The rest of this chapter is composed of a background part on the pervasive computing domain and its requirements. This is followed by our vision of the home environment, and how devices, networks and applications should be organized. Then, we describe our work on proposing a home gateway to support the home applications execution. This chapter is then concluded by describing the lessons learned from our experience on providing high level services in the home environment and the perspectives of this work.

2. Pervasive Computing

The pervasive or ubiquitous computing domain corresponds to a model of computing where the user interacts naturally with his environment. The proposed model consists of using the objects in the environment as a way of interaction between the computing system and the user. It was introduced for the first time in 1991 by Mark Weiser in articles [1, 2] presenting his vision of the 21st century computing.

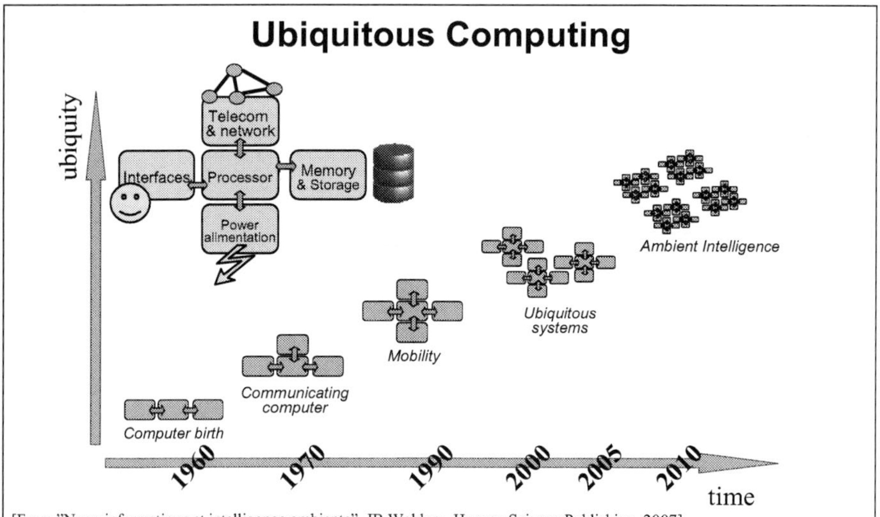

[From "Nano-informatique et intelligence ambiante", JB Waldner, Hermes Science Publishing. 2007]

Fig. 1. Computer evolutions since 1960

This new trend of computing is a natural corollary to the evolution of communicating devices. Fig. 1 represents the evolution of computing devices from the 1960's to our days highlighting the main characteristics that lead to the development of pervasive computing. Devices involved in pervasive computing systems must have the following characteristics:

- **Miniature**. As devices must be able to naturally blend or disappear in the environment, the miniaturization of devices is necessary.
- **Communication**. The devices must be able to communicate and to interact with other equipments that are present in the environment.
- **Autonomy**. The devices must have their own source of energy in order to be autonomous.

The recent evolution of computing equipment has contributed to the creation of devices which are coherent with the vision of a pervasive world. Since the invention of computing and the first computer, vendors are engaged in a race for miniaturization and processing power. While in the 1950's there was one machine for multiple users, we currently have multiple machines for one user.
Another important point is the emergence and popularization of communication technologies between computing equipments. The advent of the Internet in the 1990's corresponds to a key point, forming a network of devices from all over the world, using low cost and widespread technologies for communication. In more recent years, new communication technologies have been developed allowing wireless communication (e.g. Wifi, GSM, Bluetooth and RFID). These technologies are now embedded in devices allowing them to communicate with each other thus augmenting their mobility.
The progress achieved in energy autonomy plays an important role in the emergence of this computing domain. Some computing devices are now capable of functioning for years (e.g. RFID sensor) using a simple battery. Device autonomy represents a key factor. If this autonomy proved insufficient, the domain of pervasive computing would lose its interest because its main actors (i.e. the mobile devices) would not have the required features.
In the course of a day, a user is successively immersed in different pervasive environments and has the possibility of taking advantage of each one of these environments.
The pervasive environment in a vehicle is particularly interesting. An example scenario would consist in automatically regulating the destinations in the vehicle's GPS based on the information extracted from the schedule of the day stored in the PDA. The favorite songs of the driver can be downloaded from the MP3 player and played in the vehicle's stereo. Either at the expected time/mileage when the car needs a check up or due to an upcoming trip, the vehicle can inform the user and propose dates for an appointment in the closest auto repairer, taking into accounts the time constraints of the vehicle owner and the auto repairer schedule and availability.
In the work environment several scenarios are envisioned. Devices used in these environments differ according to the nature of the work. The devices may be handheld telephones, PDAs, printers, copiers, desktops, server computers, video projectors, or even coffee machines.
The environment of a restaurant or a bar can also offer interesting pervasive services. For example, the customer may get an interactive restaurant menu where photos, information and pictures of the dishes can be found. It is also possible to directly place an order by using the PDA which would then automatically calculate the bill.
These different environments have the objective of assisting the user in their daily life. A pervasive environment frees users from certain constraints of their daily lives by offering different services using objects from the day to day environment. The services correspond to the applications that execute on top of the host infrastructure of the pervasive environment. These services could be extremely simple, such as an electronic agenda, which interacts with single equipment, or more complex, such as a system which enables energy savings in a home, requiring the interaction with multiple devices.

3. The Home Environment

The home environment corresponds to a subset of pervasive applications that deals with the automation of home devices control. To that end, electronic devices present in the environment have the ability to communicate. The communication protocols used differ according to the type of equipment. An automated home typically allows the control of room luminosity, opening and closing of shutters, heating and air conditioning, or multimedia systems.

Home applications main objective is the comfort and simplification of the daily life of residents, and home support of elderly or convalescents. Several application areas are covered ranging from applications for the supervision of convalescents at home, to home theatre applications and energy consumption control.

3.1 Equipments

The equipment typically involved in pervasive homes include shutters, lighting devices, appliances such as coffee machines, refrigerators, or washing machines, televisions, or multimedia servers, which can all be controlled remotely. Micro-informatics equipment, such as computers, PDAs, monitors, are also part of the home automation sphere. The communication equipment is also essential in this type of system: Internet access points, routers and mobile phones allow access to information technology that may be outside of the home. Also, controllable electrical equipment such as remote electrical plugs, are a centerpiece of these new environments.

These devices must be able to be remotely controlled by the applications that coordinate their actions. Therefore, devices must provide protocols for allowing such type of communication. Nowadays there are more than fifty communication protocols, workgroups and standardizations of protocols for home communication. Among the most popular we can find X10, KNX, EIB, INSTEON, Zigbee, Bluetooth, UPnP and DPWS. These protocols allow the communication between home devices by using different transmission mediums: dedicated communication cables, communication over power lines or transmissions by radio frequency. Such protocols do not provide the same functionalities and hence are not used by the same types of devices. X10, KNX, EIB, INSTEON and Zigbee are dedicated to small devices like light dimmers and shutters. Their communication capacity is extremely limited, but their energy consumption is also limited. However, these protocols only handle communication and are not able to provide device discovery.

UPnP [3], Bluetooth and DPWS [4] are higher level protocols which handle not only communication but also device discovery. Bluetooth is a standard originally conceived for allowing wireless communication between computers and their peripherals. The goal of UPnP and DPWS is to allow peripheral devices to easily connect to each other and to simplify the implementation of home networks (e.g. file sharing, communication and entertainment) and enterprises. UPnP allows such capabilities by defining its communication and discovery protocols on top of existing Internet communication standards such as HTTP, while DPWS basically use the standard defined by Web Services.

3.2 Applications

This section presents three applications that are representative of the field of home computing. The first example is an application keeping convalescents or elderly at home.

This application requires the use of sensors specific to this field such as cardiac monitors, blood pressure gauges, sensors and video surveillance cameras. These devices collect data about the patient, and send it to an application that handles it. If a problem appears, an alarm is automatically triggered to call for help. In normal operation conditions, reports are built by the application and sent regularly to the hospital or to the doctors.
A second example is an application that manages a multimedia entertainment system from the house. The application offers the residents a set of movies from the movie library, and also offers the rental or purchase of movies from Internet service providers. The film is then automatically projected on the screen of the room in which the user is located and the atmosphere associated with the action of watching a video is applied: closing the shutters, and dimming the lights. The ring of the phone is automatically cut off. Instead, the user will have a discreet warning on their display screen on any incoming calls. He may decide to ignore it or to take the calls and the film will be automatically paused.
The last example of application consists of managing the security and minimizing energy consumption of the house in the absence of the user. When the user leaves his home, the lights are turned off, the temperature of rooms is automatically lowered, the alarm system connected and the shutters closed. In case of prolonged absence, a simulation of presence is triggered and regularly opens the flaps and lighting lamps according to the habits of the user. When the user returns home, the system is unplugged, the alarm is stopped; lighting and flaps operate according to the brightness outside.

Based on our experience, we have classified these applications into three categories based on different life cycles of these applications (see Fig. 2).

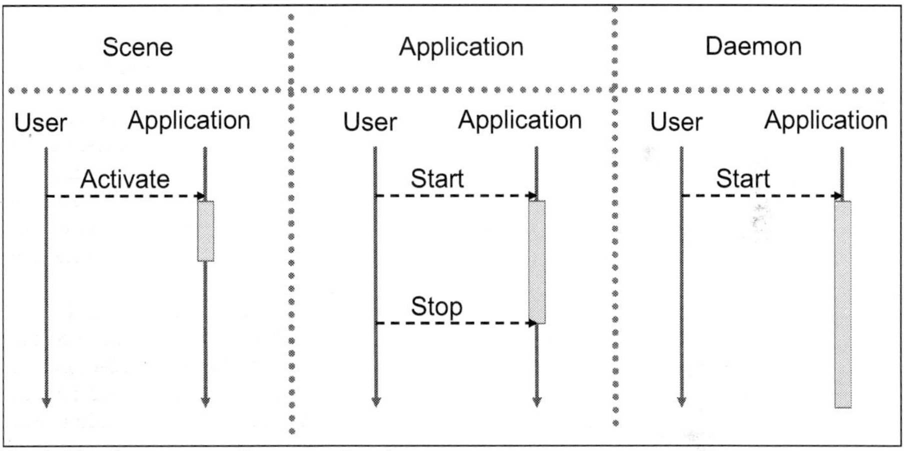

Fig. 2. The three types of home applications

A particular type of application is called scene. These applications consist of running a series of preconfigured actions at the time they are triggered. This type of application executes all of these actions and ends. For example, when the user wakes up, the application will open all shutters of the house, raise the temperature of the bathroom, heat the coffee machine and turn on the television on the favorite news channel of the user.

Another type of application, which will be called instanciable or simply application, runs for a delimited period in time. These applications have the particularity of being started and then stopped. For example, an application is activated at the time the user leaves the house. This application consists of managing the security while minimizing the energy consumed by the house. The application will turn off all the lights in the house, close all shutters, lower the heating, connect the alarm system. When the user comes back home, the application is stopped and disconnected, the alarm system is turned off and according to the circumstances, it may reopen the shutters or turn on the lights.
The last type of application is called daemon. This type of application is started and then runs continuously. For example, an application for the surveillance of patients in their home will collect medical data about the patient and then send daily reports to the doctor. Such applications are executed and never stopped.

3.3 Challenges

This domain of computing encompasses a large number of applications particularly useful to help people in their daily life. The main challenge of pervasive computing is to provide a coherent pervasive environment, providing useful services and applications, involving a set of heterogeneous, distributed and dynamic equipments and software, communicating across different protocols. In this context, several characteristics specific to the field of pervasive computing makes this area attractive in terms of industry and users, while raising difficult scientific problems for the development and management of these systems.
Distribution. The devices are an integral part of the environment. They are scattered in the physical environment and are accessible through different protocols that can use cable or wireless technologies. Applications using the capabilities of such equipment do not necessarily run on the considered equipments and are therefore distributed.
Heterogeneity. There are currently a large number of software technologies and communication protocols for the field of pervasive computing. Today there are no plans on how to reach a consensus on a common and uniform communication protocol. More than fifty protocols, working groups and specification effort are already available for home networks. The standardization of communication protocols is not possible because the devices can be of very different nature, having an impact on the communication protocols used. For example, a lamp communicates through a very simple protocol, while a PDA or a media server can use more complex communication protocols, for example considering security. In addition, manufacturers supplying equipment and protocols have no strategic interest in this type of uniform protocol, since they would lose control of their equipment.
Dynamism. The availability of equipment in a pervasive environment is much more volatile than in other areas of computing. This problem is caused by several factors including: 1) users move freely and frequently changing their location having an impact on the position of equipment they carry; 2) users can voluntarily turn on and off the devices or they may inadvertently run out of battery; 3) users and providers may periodically update the deployed services.
Multiple provider. The devices in a pervasive environment generally come from different vendors. In addition, applications deployed and running on such equipment can be delivered by other suppliers. In this context, some applications will be established through collaboration between different providers involving the creation of applications with several administration authorities. It is envisaged that the equipment vendors and service providers

want to keep some control over their devices and software and thus limit the access to external entities.

Scalability. The number of devices present in a pervasive environment can be very important. This creates a problem of scalability in applications running in this type of environment. Thus pervasive systems must be capable of handling a large number of equipment that is also dynamic.

Security. Security is a key role in building pervasive environments. Indeed, such open systems allow people in the environment to have access to the computing system. However, access to certain devices or personal data must be highly secured. The applications running on this type of system should guarantee the confidentiality and integrity of data. Access to private pervasive systems such as automated homes or cars must also include access control to ensure, for example, that a thief will not be able to disconnect the alarm system by connecting the pervasive system.

Auto-adaptation. In addition to the dynamism of software and equipments, pervasive systems are constantly faced with the evolution in their execution context. These evolutions may include changes in behavior, location, mood and habits of users, as well as changes in behavior or availability of other software. The applications running on this type of environment must be able to adapt to these changes and develop strategies to address the various events that may occur during their execution.

Simplicity of use. Finally, an essential characteristic of a pervasive system is the simplicity of use and management. Indeed, this type of system is intended to be used by users who have no knowledge in informatics. As a consequence, pervasive environments must be accessible to any human being, and even transparent. The purpose of pervasive computing is to make the devices disappear from the environment. This means that the access interface to pervasive systems must be easy to use and the applications running in these systems must be capable of adapting to different events that can intervene to maintain services usable in all circumstances.

4. Architecture of the home environment

One of the major challenges for creating an intelligent house is the design of an open infrastructure for implementing home automation applications. Equipment manufacturers and Internet operators have proposed different architectures.

4.1 Current architecture

Most current systems are based on architecture similar to the one shown in Fig. 3, in which a web server is connected to an Internet gateway using the HTTP protocol or other protocols over IP. The objective of this Internet gateway is to bridge the local network connecting the various equipments in the house and Internet. The home automation services are implemented as distributed applications running on the Internet server and in house equipments. Several gateways provided by different actors (telecom operator and electricity suppliers, home automation equipment suppliers) may be present in one house. Although this architecture allows the implementation of pervasive services for home, it also suffers from some limitations. Most treatments and coordination are made on the server side, affecting the scalability and flexibility:

- The server must handle the additional load when multiple gateways are added or when the number of connected devices in a home increases. The amount of information transmitted between the Internet gateway and the server increases proportionally with the number of equipments in the house.
- The server must know each new equipment introduced into a home to allow the dynamic evolution of services. Thus, the life cycle of equipment must be managed manually because the automatic detection of equipments availability is not feasible in a network of this scale.

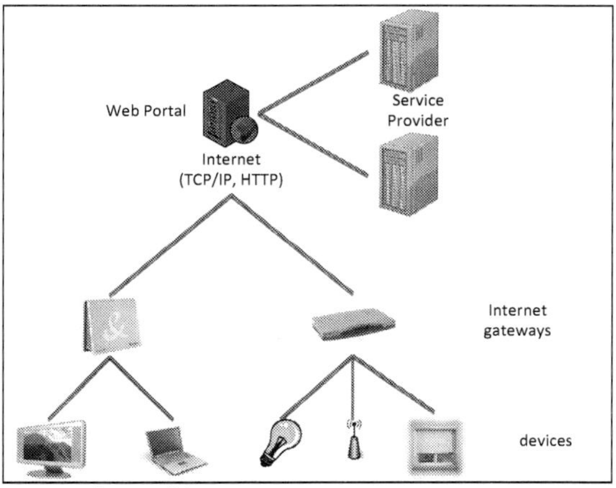

Fig. 3. Usual home computing architecture

4.2 Architecture for a home environment

To overcome the various limitations of commonly used architectures in this domain, we have proposed an innovative architecture [5] for home automation environments (Fig. 4). This work was partially supported by the European ITEA ANSO project. The home environments consist of various equipments from different vendors. We have classified equipments into three categories:

- The electronic equipments available in the house (for example controllable shutters or lamps) provide basic services to sense and act on the environment. Such equipments can be static as lamps or shutters, or may appear and disappear dynamically (such as cellular phones).
- Gateways provide an execution infrastructure for running high-level services or applications aggregating the behavior of basic services provided by the previous equipment.
- Interacting devices (such as televisions, mobile phones, or PDAs) allows users to interact with the system and potentially to manage it. Inhabitants will use them interchangeably to interact with their environment (depending on their habits and their current context).

The proposed architecture is illustrated in Fig. 4. This architecture provides a middleware as a corner stone of the home environment. This middleware provides a substrate for running

residential applications coordinating the behavior of different devices and ensuring a natural interaction, sometimes invisible, with the user. For example, an application running on this middleware can coordinate the behavior of specific equipments such as shutters, air-conditioning or lighting systems.
This middleware provides an execution environment which may be distributed across several gateways. Each gateway is usually materialized in the form of a box embedding a computer with reduced electricity consumption. These boxes generally include a set of physical communication facilities to enable interactions with actual devices. One of these boxes enables Internet access and act as an Internet provider for our middleware.

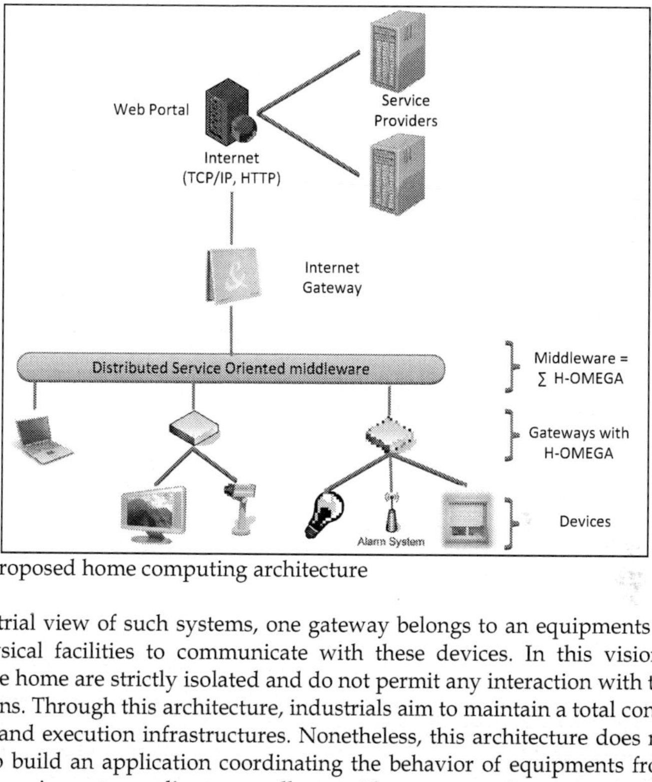

Fig. 4. Our proposed home computing architecture

In the industrial view of such systems, one gateway belongs to an equipments vendor and embeds physical facilities to communicate with these devices. In this vision, gateways present in the home are strictly isolated and do not permit any interaction with their devices or applications. Through this architecture, industrials aim to maintain a total control on their equipments and execution infrastructures. Nonetheless, this architecture does not offer the possibility to build an application coordinating the behavior of equipments from different vendors. As equipment suppliers generally provide one type of equipments, the isolation principle advocates by this architecture quickly becomes a limitation for designing innovative applications. For example, Schneider Electric is specialized on providing controllable lighting systems, and shutters, while Sony is specialized on providing multimedia systems. Thus, it is not possible to implement the multimedia entertainment system proposed in the section 3.2.
Our proposition is to take into consideration the industrial need to keep control on their infrastructure and equipments while allowing the construction of distributed applications involving pieces of software and physical devices from several gateways. Thus, our design

proposes to install on each gateway an application server dedicated to residential computing technology called H-OMEGA [5]. This application server provides an infrastructure for running residential applications. The set of application servers installed on each platform forms our middleware. Each gateway is then able to communicate with each other thanks to our middleware. Thus an application may be distributed across the different gateways present in the house and seamlessly coordinate the behavior of a set of electronic devices connected to different gateways.
H-OMEGA allows a uniform access to in-house equipments connected to the considered gateway. H-OMEGA also provides a way for applications to interact with remote services such as web services. The last feature offered by our applications server is the ability to provide an integrated and portable human interface for controlling the home system. This interface is either available from within the home, or remotely.

Our home environment, including both gateways and equipments, follows principles advocated by the Service-Oriented Computing. Service-Oriented Computing (SOC) [6, 7] is a relatively new trend in software engineering whereby services can be supplied by multiple service providers and feature various implementations. At runtime, a service consumer is able to invoke a service by relying only on a service specification, which specify both functional (service interface) and non-functional (QoS) part, while not referring to the service implementation. An important consequence of this interaction pattern is that SOC technologies support dynamic service discovery and lazy inter-service binding. Such characteristics are essential when building applications with strong adaptability requirements, such as pervasive and residential applications.
We propose to build smart home applications as service-oriented applications. The H-OMEGA application server is based on service-oriented architecture and interactions with remote devices follow the service-oriented pattern. The use of this technology presents several advantages. As previously stated, this technology allows the loose-coupling between different actors which allows the use of other services without having detailed knowledge of their implementation or the interaction protocol used to communicate with their equipment.
We propose to reify each device feature as a service on our middleware to provide an uniform access to electronic devices. This technique addresses the problem of heterogeneity of communication protocols between the various equipments. In addition, the use of the service-oriented components approach provides a natural support for dynamic applications such as the ones found in a house. Indeed, residential applications have to interact with equipments accessible through services, which may be intermittently available. Finally the use of such technology respect the vision of the existing protocols for home: UPnP and DPWS which propose a service-oriented approach. The integration of devices that do not comply with a service-oriented approach is made through the use of a third party mechanism which makes the link between our service-oriented gateway and different equipments.
The proposed application server also allows service providers to deploy, update and remove services and applications remotely. Suppliers can thus control from their own premises all applications and services deployed in all houses.

5. Our Residential Gateway Proposition

5.1 Architectural view

The architecture of our residential gateway, H-OMEGA is illustrated in Fig. 5. This architecture is based on three basic elements used to simplify the design, implementation, development and administration of residential applications. The main elements of this architecture are:

- An infrastructure for service-oriented execution,
- A remote service manager (including equipments, services offered by other gateways and web services),
- A set of facilities or commonly used services to develop this type of application.

The rèmote service manager can manage both the available devices in the environment of the gateway, the services offered by other H-OMEGA gateways presents in the home and remote software services from outside the home. The role of this entity is specifically to manage the lifecycle of services acting as proxy for remote services either offered by remote equipments, or remote gateways. These local representatives are able to interact directly with the remote service. They follow the life cycle of their corresponding remote service. The role of the manager is to ensure a coherent behavior of these proxies. Applications on our framework have the possibility to transparently use remote services or features from remote devices through their local representatives.

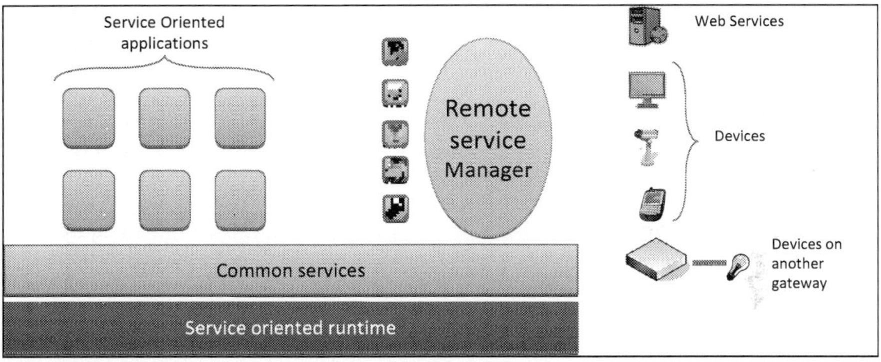

Fig. 5. H-OMEGA application server architecture

The commonly used services in applications are provided by the framework. The applications running on the framework have access to these services. The goal of creating these services is to free developers of applications from this tedious, repetitive and sometimes complex development. The use of these services helps to reduce the bugs in this type of application, because these services are developed once and widely tested. Our framework currently provides:

- A persistence manager to enable applications to store and retrieve persistent data;
- A tasks scheduler for repetitive or delayed tasks;
- An event-based communications infrastructure for enabling asynchronous communications;

- A remote administration module to easily manage deployed residential applications from the vendor premise.

The service-oriented infrastructure allows the design of residential applications with the benefits associated with this type of infrastructure. Applications can be opportunistically bound to services provided on the gateway. The services available to applications through this mechanism include the services provided by other applications, equipments and remote services accessible through local proxies and the common services provided by the platform.

5.2 Implementation

The Fig. 6 shows the stack of technologies used to develop our applications server. Our framework provides a Java-based environment to develop residential applications. It is based on service-oriented technology called OSGi [8] which is a service-oriented architecture featuring management facilities. On top of this technology, we use iPOJO [9]: a service-oriented component runtime that aims to simplify the development of service-oriented applications.

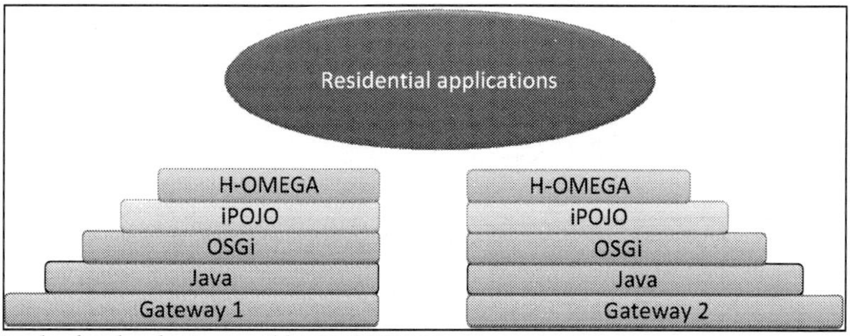

Fig. 6. Stack of technologies used by H-OMEGA

iPOJO is a service-oriented component runtime that aims to simplify the development of applications on top of OSGi SOC Platforms. iPOJO allows the straightforward development of application logic based on Plain Old Java Objects (POJO). iPOJO subsequently injects non-functional facilities into the application components, as necessary. Such facilities include service provisioning, service dependency and lifecycle management. In addition to providing a reusable set of non-functional capabilities, iPOJO is seamlessly extensible to include new middleware functionalities.

The iPOJO framework merges the advantages of components with service-oriented paradigms. Specifically, iPOJO application functionalities are implemented following the component orientation paradigm. Each component is fully encapsulated, self-sufficient, and provides server and client interfaces exposing its functionalities and dependencies, respectively. As many component-oriented platforms (e.g. Java EE and .NET), iPOJO separates a component's application-specific business logic from its application independent functions. As such, iPOJO components consist of a component implementation that is managed by a reusable container (Fig. 7).

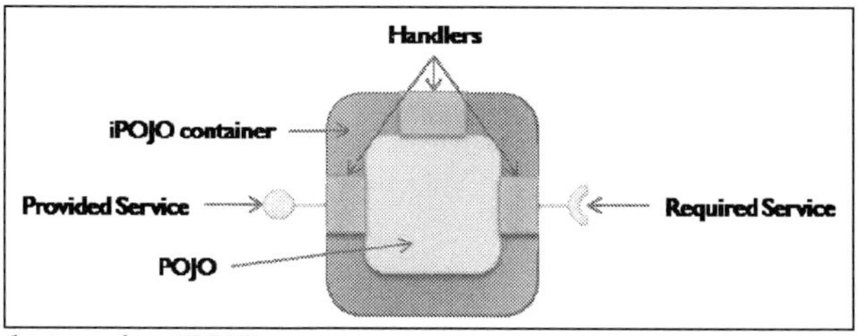

Fig. 7. Internal design of an iPOJO component

iPOJO containers provide common middleware functionalities to the component implementations they manage (e.g. distributed communication and lifecycle management). Each component container can be configured with a different set of middleware services, implemented as "handlers". Once an iPOJO component is deployed, its provided functions are published and made available as services, in conformance with the SOC paradigm. In order for a component's services to become valid, all the component's dependencies must be resolved. For this purpose, available services corresponding to a component's required (or client) interfaces must be found and available.

The use of iPOJO allows us to benefit from all the facilities provided by this technology, particularly the dependencies manager which automatically deals with the dynamic availability of services (specifically services provided by mobile and remote devices). In addition, the extensibility feature of iPOJO enables the specialization of the environment for the residential application needs. We thus have developed handlers to simplify access to commonly used features of our framework:

- A handler to describe the automatic planning of repetitive or delayed actions. This handler (called cron handler) uses the scheduler services provided by our framework.
- A handler to automatically save and restore the state of a service. This handler (called persistency handler) uses the persistence service provided by H-OMEGA.
- A handler to simplify the reception and sending of asynchronous messages. This handler (called Event Admin handler) uses the event-based communication infrastructure provided by OSGi.
- A handler to describe the provisioning of administration features of a service. This handler (called JMX handler) uses the JMX standard provided by the JVM to offers this functionality.

An application developer using H-OMEGA will thus have an easy access to all these features.

6. Examples

This work has been validated as part of the ANSO European ITEA project. The middleware presented in this chapter has been used as a basis of the final demonstrator of the project. ANSO means Autonomic Network for SOHO, where SOHO is used for Small Office Home Office. The objective of this project was to develop an open source platform, intelligent and reliable for different home automation environments to greatly accelerate the development of new services in this context and to allow their compositions in innovative applications for increase the use of services for the digital home in Europe.

In this context, we have developed several home scenarios to demonstrate the interest of our framework. The applications developed are presented in section 3.2.

The first application is a home hospitalization application to help maintain elderly or convalescents at home. Based on fall detectors and blood pressure sensors, our application constantly monitors the considered person. These data are processed through complex analyzers to detect irregularities or unexpected behaviors. In such cases, an alarm is sent to the closest hospital emergency. In normal operational condition, this application continuously stores information on the patient health and builds reports which are regularly sent to the doctor in charge.

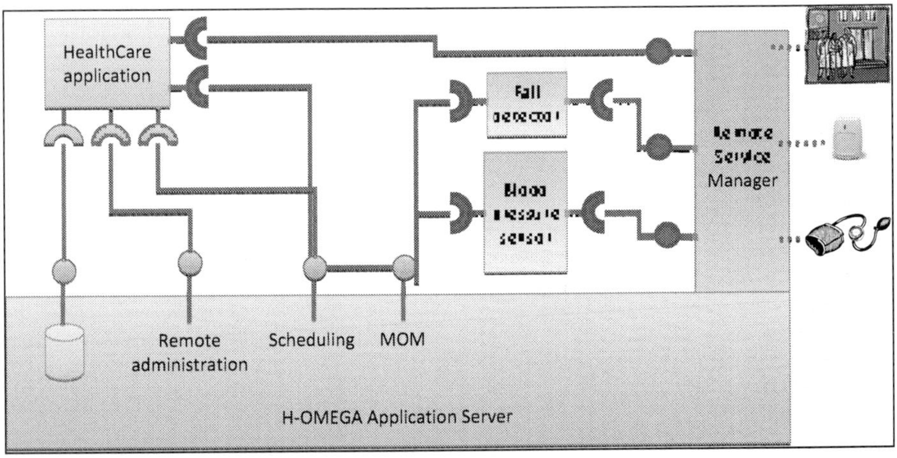

Fig. 8. Home hospitalization application

This application has been designed using the various features of H-OMEGA. As, we do not have access to the real sensors, this application has been built using simulators of the real sensors. To keep the simulation close to the real sensors, we have developed and executed these simulated sensors on a remote computer, and we have used standard protocols such as UPnP to remotely discover and access them. Thanks to our remote service manager, proxies of these sensors are automatically installed on the gateway. All data are transmitted through an event-based communication system to a service in charge of performing anomalies recognition. Data are also stored on a persistent support through the persistency service. The application uses the facility of the automatic planning of repetitive actions to plan the creation of a daily report. Finally, this application uses the remote administration feature to provide a way for the hospital to remotely tune the thresholds of the anomalies detector in order to suit with the patient health evolution.

The second application is a home multimedia entertainment application using standard UPnP media server and renderer devices. In this application, we do not simulate any devices. This application aims at providing a multimedia experience to the user seamlessly integrating several multimedia devices and the shutter and lighting system of the home.

First the user chooses a media to listen or view, and then the system uses the maximum of its capacity to maximize the user comfort. The media follows the user while he is moving throughout the house, and the suitable ambiance for watching media is also set in each visited room. This application mainly benefit from the facilities provided by iPOJO to manage the dependency between the media controller service and the available media renderer in the home. Thus, the application is able to view the media in the room where the user is located. This application is distributed across two gateways: one belonging to the multimedia vendors, the other belonging to the vendor of the shutter and lighting systems. The application mainly runs on the multimedia gateway, but uses the feature of our middleware to access the lighting services on the other gateway.

The third application is an application aiming at minimizing the energy consumption of the house while maximizing the security when inhabitants are away. This application is in charge of running the alarm system, closing shutter and turning off all lights when inhabitants leave the home. If the inhabitants' absence last more than one day, this application launches a service in charge of simulating the presence. This last service makes extensive use of the planning feature offered by our middleware to simulate the inhabitants' usual actions, such as closing shutters, turning off lights in different rooms, etc.

7. Conclusion

Developing correct and maintainable pervasive services is a real challenge today. It is clear that most techniques currently available are not mature, hard to master and, consequently, raise major challenges for the major players of the market.

We believe that two important aspects have to be improved: development environments and runtime environments for pervasive services. In this paper, we have presented recent developments in the area of service-oriented home gateways.

This chapter mainly focuses on the description of our work on a runtime addressing the main limitations of current approach: dealing with a growing number of homes and dealing with heterogeneous mobile devices. The design of our residential application server also respect the industrials main will to keep the control on their own equipments, while encompassing the main limitations of the traditional approach: entirely isolated gateways.

The work described has been implemented on top of an open source project called iPOJO (available as an Apache Felix subproject) and is currently available as an open source project on http://ligforge.imag.fr/projects/homega/. This work has been validated in the ITEA ANSO project and through the creation of several applications validating the usefulness of our framework.

This work, on providing an open infrastructure to enable the development and execution of home applications seamlessly integrating heterogeneous and mobile devices, open several research perspectives. We are currently working on adding autonomic features to home applications in order to reduce the maintenance cost of such applications [10]. This work aims at providing architecture and its corresponding runtime to support the creation of self-configuring, self-optimizing and self-repairing applications.

8. References

[1] Mark Weiser, "The computer for the 21st century", Scientific American, 265(3):66-75, September 1991.

[2] A. Ferscha, "Pervasive computing and communications", Beyond The Horizon Thematic Group, IST, 2005 (http://www.cordis.lu/ist/fet/id.htm).

[3] UPnP Plug and Play Forum, "About the UPnP Plug adn Play Forum," in http://www.upnp.org/, 1999.

[4] E. Zeeb, A. Bobek, H. Bonn, and F. Golatowski, "Lessons learned from implementing the Devices Profile for Web Services," in Inaugural IEEE-IES Digital EcoSystems and Technologies Conference (DEST '07) 2007.

[5] C. Escoffier, J. Bourcier, P. Lalanda, J. Yu, "Towards a home application server" 5th IEEE Consumer Communications and Networking Conference (CCNC'08), January 2008.

[6] M. N. Huns and M. P. Singh. Service-Oriented Computing: Key Concepts and Principles. IEEE Internet Computing, vol. 9: pages 75–81, Jan./Feb. 2005.

[7] M. P. Papazoglou and D. Georgakopoulos. Service-oriented computing. Commun. ACM, 46(10):24–28, 2003.

[8] OSGi Alliance. "OSGi Service Platform Core Specification Release 4" http://www.osgi.org, August 2005.

[9] C. Escoffier, R. S. Hall, P. Lalanda, "An Extensible Service-Oriented Component Framework", IEEE Service Computing Conference, 2007.

[10] P. Lalanda and J. Bourcier, "Towards autonomic residential gateways", IEEE International Conference on Pervasive Services (ICPS 2006), June 2006.

2

Integrated Wireless Technologies for Smart Homes Applications

Mahmoud A. Al-Qutayri and Jeedella S. Jeedella
Khalifa University of Science, Technology, and Research
United Arab Emirates

1. Introduction

Due to the rapid advances in wireless communication and information technologies it is now possible to embed various levels of smartness in the home. These smart homes are ones that can interact intelligently with their inhibitors to provide comfort and safe living. This interaction may range from simple control of ambient temperature to context-aware and mobile agent based services. An example of that is delivery of particular information content based on the smart home inhibitor location inside the home and the activities that he or she is engaged with.

Wireless networks and sensors are seen to play an increasingly important role as key enablers in emerging pervasive computing technologies that are required for the realization of smart homes. The wide spread of wireless networks in our daily life is enabled by the communication standards such as WiFi, Bluetooth, Zigbee, RFID, and cellular technologies. A combination of these standards is envisaged to be used to construct the smart home. Effectively all wireless technologies that can support some form of remote data transfer, sensing and control are candidates for inclusion in the smart home portfolio.

A top level architecture of a smart home is illustrated in Fig. 1 (Al-Qutayri & Jeedella, 2010). It includes a server/gateway/router that can be used as the central point of connectivity for devices within the home as well as allowing connectivity to the outside world. The setup also includes smart sensors as well as appliances that have either wired or wireless connectivity. Communicating with the smart home from the outside can be done using one or a combination of the following external networks such as phone lines, xDSL lines, cable TV, GSM and power line networks.

Following this introduction, Section 2 reviews some of the major wireless communication technologies that can be used to realize a smart home. Section 3 describes briefly some of the smart sensors that can be used. Section 4 describes the major computing technologies, including ubiquitous and pervasive, context-aware, and agents computing which introduce smartness aspects to the home. Section 5 details some of the application that can be realized with smart homes. Section 6 describes the design and implementation of some smart home systems. In particular, the end-to-end realization of a secure system that enables monitoring and control of various devices with multiple levels of settings is described and the results achieved are presented. Then Section 7 presents some conclusions and possible future directions in the area of smart homes.

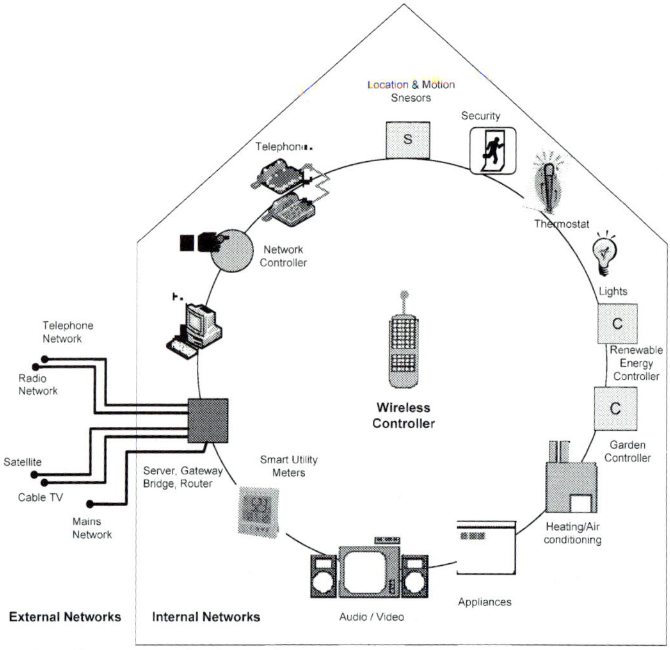

Fig. 1. Top Level Architecture of a Smart home

2. Wireless Technologies

The wireless technology standards are everywhere. Bluetooth, Zigbee, RFID, WiFi, and cellular technologies are the most well known standards. A combination of these standards is envisaged to be used to construct the smart home. Effectively all wireless technologies that can support some form of remote data transfer, sensing and control are candidates for inclusion in the smart home portfolio. This section discusses some of these key wireless technologies.

2.1 Bluetooth

Bluetooth is a universal radio interface that enables various electronic devices, including mobile phones, sensors… etc, to communicate wirelessly through a short range radio connection (Thompson et al., 2008). The introduction of this technology eliminated the requirement for wired connections, eased the connectivity process between devices, and enabled the formation of personal networks. The pervasiveness of Bluetooth enabled electronic devices is enabling ubiquitous connectivity and hence allowing the development of many applications.

A Bluetooth device uses a license-free frequency band at 2.45 GHz. This band is also known as the Industrial-Scientific-Medical (ISM) band and has a range of 2.4 GHz to 2.4835 GHz. As this band is a free one, and hence gets used by other applications such as cordless phones, Bluetooth radio transceivers use frequency-hopping spread-spectrum to avoid interference. Depending on the Bluetooth class, the communication range varies from 1 meter for Class 3 to 100 meters for Class 1. The most common range is 10 meters for Class 2. The data rate of devices in a Bluetooth network varies from 1 Mbps to 24 Mbps. In a Bluetooth network, there are two types of devices: a slave and a master. Each Bluetooth device has the ability to be either a slave or a master or both at the same time. In general, a Bluetooth network consists of small subnets or piconets. A piconet is formed by two or more connected devices sharing the same channel. In every piconet, there is only one master and up to 7 slaves. The communication between the slaves goes all time through the master. When two or more piconets are connected they form a scatternet. The connection between piconets can be done by having a device in common. This device may be a slave in one piconet and a master in another piconet as shown in Fig. 2. The protocol stack of Bluetooth is depicted in Fig. 3 and the details of its functionality are given in (Labiod et al., 2007).

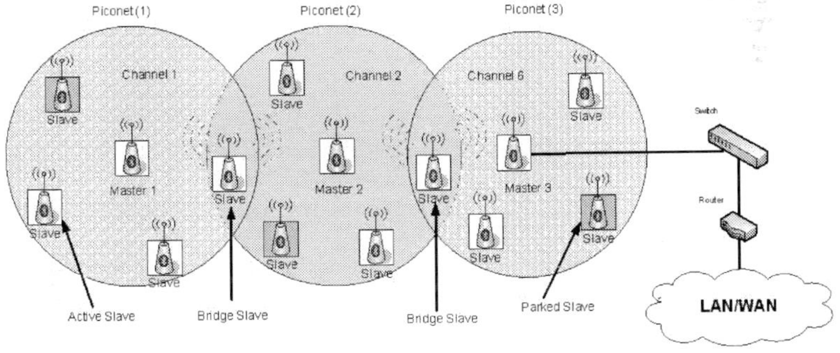

Fig. 2. An example of Bluetooth scatternet

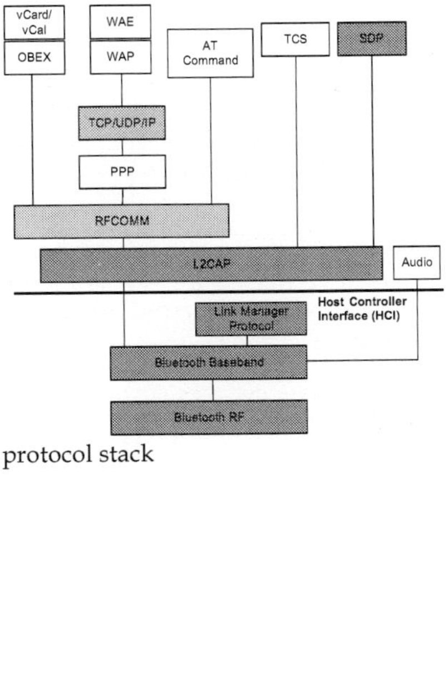

Fig. 3. General Bluetooth protocol stack

Smart homes can benefit from Bluetooth technology in a variety of ways. One possibility is to embed appliances with Bluetooth radio transceivers and use that technology to communicate with a home server that is accessible by the user. This enables monitoring and control operations to be conducted by the user. Another possible application is the establishments of Bluetooth enabled sensor networks that can track the well being of people with disabilities (Leopold et al., 2003).

The challenges that face the use of Bluetooth in a smart home environment are similar to those facing the technology in other environments. A primary concern of the use of Bluetooth is its security vulnerability. It has been shown that the security of Bluetooth devices can be compromised by adversaries. A number of solutions have been proposed in the literature to harden security and privacy of Bluetooth based communication (Carettoni et al., 2007).

2.2 ZigBee (IEEE 802.15.4)

IEEE 802.15.4 standard is a low cost low power wireless communication standard for Personal Area Network (PAN). The low cost makes it suitable for remote control and monitoring applications. The low power makes it suitable to operate on batteries for long life. It reduces the cost of hardware and consuming power by lowering its data rate. The specifications define only the lowest two layer of the OSI networking reference model: the physical and Media Access Control (MAC) layers. The data rate, operating frequency, and network size are defined by the standard. The achieved data rate between IEEE 802.15.4 compliant devices varies from 250 kbit/s to 20kb/s depending on the distance between devices and the transmission power. These devices may operate in one of the following three RF bands: 868 MHz (Europe), 915 MHz (North America), and 2400 MHz (worldwide). The 2.4 GMhz band is used more often than the other bands since it is available worldwide for unlicensed operation. In addition to that, the performance of products developed for that band is better when compared to the other bands with respect to data rate. The size of the network is not limited by the standard. However, network address are stored and sent using 16 bit or 64 bit numbers, which limits the network size to 2^{64} devices.

IEEE 802.15.4 standard defines Star, Cluster Tree and Mesh networks as possible topologies for the wireless network as shown

Fig. 4. However, mesh networks enable high levels of reliability and longer coverage range by providing more than one path through the network for any wireless link. Note that in any ZigBee network there are three types of ZigBee devices (Gislason, 2008):

- **PAN coordinator**: There is only one coordinator in a network that is responsible for starting the network, binding together devices. Also it routes data between different devices. It is a Full Function Device (FFD) and it is usually mains powered device.

- **A router**: It cannot start the network however it scans a network to join it. Once it is in the network it can route data between Reduced Function Devices (RFD). It is a FFD and it is usually mains powered device.

- **An end device**: It cannot start a network however it scans a network to join it. It can be either a RFD or FFD and it is usually battery powered device.

The protocol stack of Zigbee defines only some functionality in layers on top of the physical and MAC layers which are defined in the IEEE 802.15.4 standard. It provides the set of programming tools for the intended market. Furthermore, ZigBee technology defines a set of applications profiles to facilitate the development and deployment of ZigBee devices from different manufacturers as shown in Fig. 5 (Gislason, 2008).

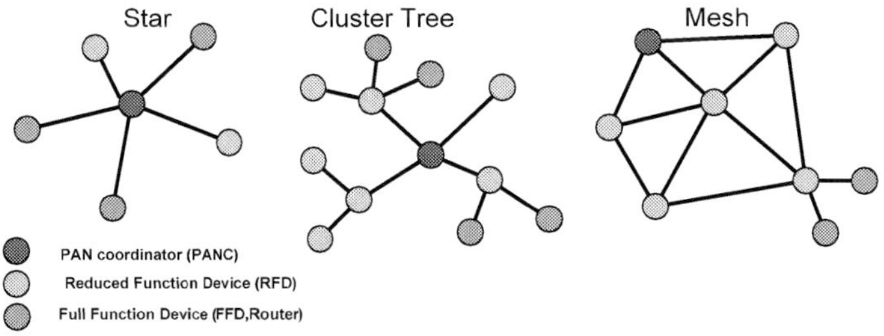

Fig. 4. Possible ZigBee networking topologies

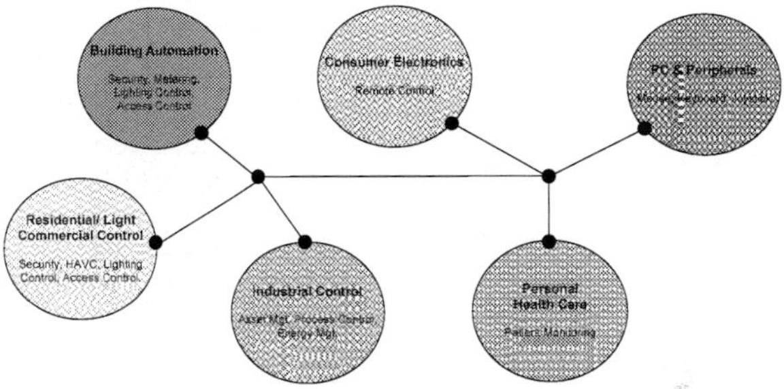

Fig. 5. ZigBee Applications Profiles

2.3 RFID

Radio Frequency Identification (RFID) describes a system that transmits the identity of an object wirelessly using radio waves (Want, 2006). It defines a RFID tag holding information about the object carrying the tag and a RFID reader. The RFID tag transmits signals containing its data when it is scanned by the reader. The RFID tag can be either active or passive where an active tag contains a battery and the passive tag does not have a battery. The passive tag uses the reader's magnetic field and converts it to DC voltage to power up its circuitry. Consequently, the passive tags are cheaper and have lower range when compared to active tags.

RFID systems can be categorized based on the used frequency ranges. The Low-Frequency (LF) systems use signals with a frequency between 124-135KHz. The High-Frequency (HF) systems use a 13.56MHz and the Ultra-High-Frequency (UHF) systems use a frequency between 860-960MHz. In general, the LF RFID systems have short reading ranges and lower system costs. In case longer reading range is required, HF RFID systems can be used however their cost is higher.

RFID systems can be used in smart homes where every single object can be connected to the Home Area Network (HAN) through a virtual wireless address and unique identifier (Darianian and Michael, 2008). This can be used to keep an updated database holding information about objects' locations. Accordingly, the smart home can be asked to provide information about a specific object that you are looking for such as your car's key or your remote control. Furthermore, RFID system can be used to track smart home occupants, where a number of studies have been reported in the literature that use RFID concept to track smart home occupants (Yamazaki, 2007). By the attachment of a RFID tag to each smart home user and the deployment of RFID readers at different places in the home, the location of each user can be identified. This information can be used to adapt services in the smart home based on each user preferences.

One of the problems of using RFID tags to track people in smart homes is that the readability of RFID tags is difficult near water or a sheet of metal. The human body consists primarily of liquid which makes it difficult to scan a RFID tag attached to human body (Juels, 2006). However, researchers are looking for new ways to improve the readability of RFID tags in these difficult environments.

2.4 GSM/GPRS

The GSM (Global System Mobile) is the technology that generated a revolution in the field of mobile communications. New generations of GSM were introduced over the past decade that includes GPRS, UMTS... etc in order to improve the transmission rates, and offer new types of services (Brand and Aghvami, 2002). The GSM which is also known as the cellular network is based on frequency reuse. To that effect a particular geographical area gets divided into cells. The size of the cell is normally dependent on the local traffic distribution and demand. A high level architectural view of GSM/GPRS is shown in Fig. 6.

The mobile wireless system such as GSM/GPRS is used to deliver both voice and data communications. One of the cost effective services that is delivered by the network and can be used for smart homes applications is the SMS (short message service). The SMS is a text message whose content can be processed using an appropriate program in order to execute commands for monitoring and control operations (Al-Qutayri et al., 2008). Such programs are normally written using Java language. The ability to use the GSM network basically means that remote access and control to a smart home is possible.

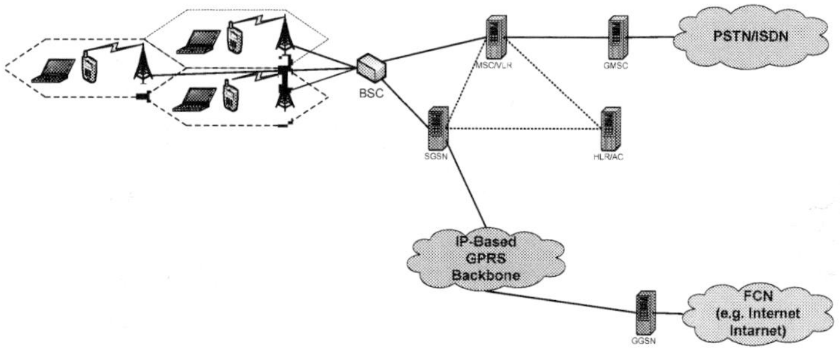

Fig. 6. GSM–GPRS System Architecture

2.5 WiFi (IEEE 802.11)

Wireless Fidelity (WiFi) is a common term that refers to the IEEE 802.11 wireless communication standard for wireless local area networks (WLAN) in the 2.4, 3.6 and 5 GHz frequency bands. Network users, when using WiFi technology, can move around without restriction and access the network from almost anywhere. Also it can provide a cost-effective network setup for hard-to-wire locations such as old buildings. Two types of devices are considered in the WiFi standard: an access point (AP) and a wireless device which could be a laptop equipped with a wireless network interface. The main function of an AP is to bridge the information between the fixed wired network and the wireless network. An AP can support up to 30 wireless devices and can cover a range of 33–50 meters indoors and up to 100 meters outdoors. The wireless devices can be possibly connected together using infrastructure topology or an ad hoc mode topology.

The infrastructure topology is sometime called an AP topology since the wireless network consists of at least an AP and a set of wireless devices. In this topology, the system is divided into basic cells, where each cell is controlled by an AP. To extend the coverage area, multiple basic cells can be used as shown in Fig. 7.

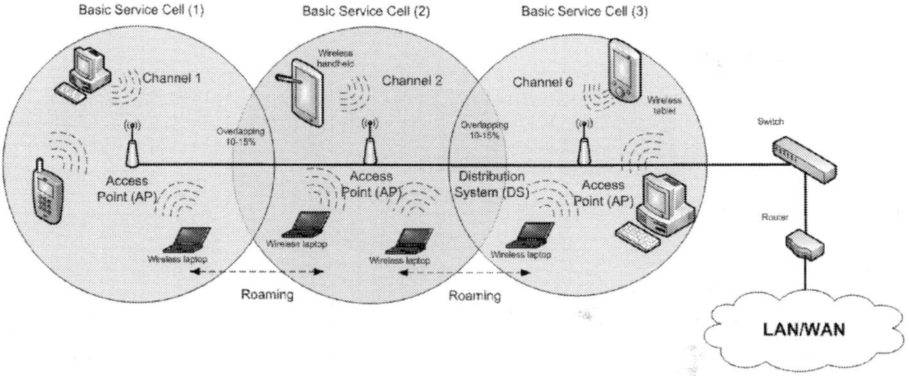

Fig. 7. A typical WLAN

In general, wireless networks should be able to reach fixed Local Area Network (LAN) services such as file servers, printers and Internet access. This is achieved by the distribution system (DS) connecting the different APs together. The connection between the APs can be done using either a cable connecting them together or using a wireless connection. The data transfer between wireless devices within a basic cell and the distribution system occur via an AP. The distribution system is responsible for transferring the data packets between various cells within the wireless network. It is also responsible for address mapping and internetworking functions. To cover an extended area, basic cells may sometimes partially overlap as shown in Fig. 7. On the other hand, the ad hoc topology represents a group of WiFi devices that have the ability to dynamically form connections with each other to create a network. This ad-hoc network does not require a connection to either an AP or to fixed network. It can grow, shrink and fragment without having to make any requests to a central authority. It is useful for setting up a wireless network quickly and easily.

IEEE 802.11 standard is similar to IEEE 802 standard that deals with LANs and metropolitan area networks (MAN). It focuses on the two lowest sub-layers of the Open System Interconnection (OSI) networking reference model. Namely, the physical layer (PHY) and a data link layer containing the MAC sublayer and the LLC and more details can be found in (Labiod et al., 2007).
The IEEE 802.11 standard has evolved over the past years where two types of systems were defined. Those operate in the band of 2.4 GHz such as IEEE 802.11b/g and those operate in the band of 5 GHz such as IEEE 802.11n. Since IEEE 802.11n standard supports high data rate approximately five times higher than the previous standard, it is expected that it will be used in consumer electronics applications, especially for streaming video in smart homes. The video signal can be displayed on the suitable display system based on the smart home inhibitors locations and preferences. Some companies such as Philips are demonstrating wireless video streaming for home entertainment system using this wireless technology. Since the existing 802.11a/b/g standards were created to serve the PC applications domain, they have substantial limitations for real-time and high bandwidth requirements from consumer electronics applications. Even though the 802.11g has a maximum data rate of 54Mb/s, in practice it achieves 20Mb/s with difficulty, especially when the signal has to penetrate walls. With the improvements in codec technologies such as MPEG4, H.264 and WMV9 the required bandwidth to stream video is reduced. However, other requirements are driving to increase the required streaming bandwidth such as high definition video, the Voice over Internet Protocol (VoIP), networked audio devices, etc.

3. Smart Sensors

A sensor is a device that converts a physical or biological quantity into electrical quantity. The measured electrical quantity should be calibrated, converted to digital format and sent to the microcontroller for further processing and control. Most of the sensors, irrespective of their types, can be included as part of a ubiquitous embedded system that has communication capabilities and backend connectivity. These types of sensors are called sometimes smart sensors. These enable software development and data analysis using embedded processing capabilities as well as sending remote processing by a computing system located at some other location. Examples of these types of integrated sensors include Particle Computer (Decker et al., 2005), Berkeley MOTES (Estrin, 2002)… etc.
The intelligent homes of the future are expected to be embedded with a network of heterogeneous low power wireless smart sensors that monitor the vast set of parameters necessary for building ambient intelligence. The smart sensors will have to function in an autonomous manner and maintain the privacy of the home inhabitants. The tasks that the network of sensors and actuators may perform in a smart environment can range from a simple one such as turning on and off the garden sprinkler system at regular intervals to supporting elderly people at home (Dengler et al., 2007).
The set of sensors and actuators that will be needed can be broadly classified as conventional and non-conventional types with reference to present state of sensors technology. The conventional sensors type includes temperature, humidity, light, motion and smoke detectors (Augusto and Nugent, 2006). A light sensor for example can be used to automatically differentiate between day light and nightfall and hence open or close the curtains using an appropriate actuator. Temperature and humidity sensors can be used in

conjunction with the heating and air-conditioning system to optimise the home atmosphere and give the same level of comfort throughout the home.
Non-conventional sensors types include location, posture, heartbeat and biosensors. These types of sensors enable monitoring various conditions including health oriented ones for elderly persons living on their own. The biosensors such as finger print, face and iris recognition can be used to grant physical access to the home. Accordingly, the house environment will be adapted to his or her needs.
In order to build the ambient intelligent environment for the smart home inhibitors different types of sensors should be deployed in the house. By the fusion of the data streams from different sensors the whereabouts of smart home users can be inferred. For example, the data coming from audiovisual sensors can be combined with the RFID tracking system to infer the location of smart home owner and his activity. In some cases the emotion of a smart home owner can be inferred using face recognition and the analysis of his or her voice.

4. Computing Technologies

The sensory information gathered by the distributed sensors in a smart environment needs to be processed in order to perform the required actions. These actions can occupy a wide spectrum that extends from simply toggling a device between on and off states to sophisticated personalized decisions. The level of sophistication and intelligence that needs to be used in interacting with devices depends on the complexity of the device and the type and level of functions it can perform. An air conditioning system for example has a reasonable level of complexity. For such a system, once it gets switched on the desired temperature and air flow levels need to be set.
Interacting with individual devices and appliances can introduce a basic level of intelligence to the home environment. However, the level of intelligence can be greatly enhanced once devices, be it simple sensors or complex appliances, can exchange information and effectively share the decision making process to offer a certain type of service to the occupant of an intelligent environment. This section reviews the major computing technologies that have been proposed in the literature to realize smart homes or more generally smart environments.

4.1 Ubiquitous and Pervasive Computing

Advances in integrated circuits design and miniaturization processes are resulting in computing devices with ever shrinking sizes. At present, such devices are being commercially produced using an approximately thirty nanometer process. This reduction in the computing devices size and cost has effectively led in recent years to the realization of the ubiquitous computing world envisaged by Mark Weizer (Weizer 1991). In that vision Weizer stated that "the most profound technologies are those that disappear, they weave themselves into the fabric of everyday life until they are indistinguishable from it."
In the ubiquitous world of computing, which was subsequently named by IBM as pervasive computing (Satyanarayanan, 2001, Garlan, 2002), the technology becomes unobtrusive as computers get pushed into the background and the focus shifts to the information and the people using the system. The major aspects of pervasive computing systems include (Haryanto, 2005):

- Computing devices are scattered everywhere in the environment or space of interest
- The devices can be interconnected and interact with each other
- The devices can work autonomously without active attention from the user
- Each device tends to be dedicated for a special task
- The devices integrate into the background in a seamless manner that makes them effectively invisible to the user.

Pervasive computing plays a major role along with other technologies, particularly computing ones, in supporting the realization of smart homes as well as the more general ambiance intelligence (Ma et al., 2005; Mann, 2005; Augusto, 2007). To that extent many researcher have attempted to use pervasive computing to introduce smartness to some aspect of the home while other built a complete smart home prototype with extensive devices and networking. The articles by (Al-Qutayri et al., 2008; Choi et al., 2005; Roduner et al, 2007) describe smart home systems in which some of the appliances are fitted with wireless sensors that enable remote monitoring and controlling of those devices. Other significant research work that resulted in smart home structures with extensive pervasive computing infrastructure include AwareHome (GATECH) (Kientz et al., 2008) and Adaptive House (University of Colorado) (Mozer, 2005) as well as other projects (Helal et al., 2005).

4.2 Context-Aware Computing

Context-aware computing and pervasive computing go hand in hand and complement each other. Having pervasive computing environment provides the infrastructure that supports advancements in context-aware computing. Conversely, the enhanced capabilities of context-aware computing are placing demand on building responsive pervasive computing environments (Baldauf, 2007; Dargie, 2009). Computing systems that are context-aware have smart or cognitive characteristics that enable them to adapt to changes in the environment and user requirements (Schaefer, 2006). This dynamic adaptation provides the system with the autonomy needed to take decisions without the user direct intervention and hence make computing an unobtrusive task that blends into the fabric of the environment. Extensive research has been conducted to date to enable the design of context-aware computing system. All such systems tend to have architecture with a middleware layer that interfaces the application to the system software. The middleware supports integration between disparate products and platforms while maintaining the integrity, reliability and robustness of the overall solution.

The application of context-aware computing in the field of smart homes continues to be of interest to many researchers. It creates the required smart features that allow flexible interaction between the user and the environment. The article by (Oh and Woo, 2005) proposes a unified ubiHome application service model that provides personalized environment to each user. Application of context-awareness in the control of smart home appliances is described in (Choi et al., 2005). The system utilizes six parameters for learning and predicting the user's preference: the pulse, the body temperature, the facial expression, the room temperature, the time and the location. The application of context-aware computing in a smart home environment to support independent living and provide health care services is the subject of extensive research given the importance of this field (Haryanto, 2005; Cook, 2006; Bricon-Souf and Newman, 2007).

4.3 Agent Based Computing

Agent based computing is a powerful technology for the development of complex software systems. It draws on knowledge from different areas including software engineering, artificial intelligence, robotic and distributed computing. An agent is characterized by autonomy, social ability, reactivity, and pro-activeness. These characteristics enable the development of proactive and intelligent applications that can interact with one another (Zambonelli et al., 2000).

A multi-agent system can be considered an ensemble of autonomous agents. The agents work independently from each other. However, as each agent tries to achieve its task it typically needs to interact with other agents as well as the surrounding environment in order to obtain information and/or services that it does not have or coordinate its activities in such a way to achieve its goals.

Multi-agent systems tend to be static and hence execute on the system on which they are running. However, by their very nature mobile agents are programs that can autonomously move from one computing machine in a network to another. The dynamic movement ability of a mobile agent enables it to move to a computing system on the network that contains the object it wants to interact with (Ilarri et al., 2008).

Agent based computing technology, be it static or mobile, lends itself well to smart homes research. The article by (Cook et al., 2003) describes the "MavHome" project whose goal is to create a home that acts as intelligent agent. The study by (Velasco et al., 2005) proposes a multi-agent based architecture which enables multimedia contents to follow the user movements throughout the smart home environment. The article by (Marsa-Maestre et al., 2008) proposes a service oriented architecture implementation based on mobile agent systems. This is then used for service personalization using mobile agents that follow the system users as they move from one location to another within a smart space.

5. Applications

The opportunities that are primarily envisaged in a home that can be classified as smart are numerous and can be classified as services or monitoring and control. The sub-sections below explore some of those smart home applications.

5.1 Appliance monitoring and control

By exploiting sensors in smart home appliances and connecting them in smart home network, they can operate in a much more sophisticated and intelligent ways. They could be controlled easily from any place in the house by switching them ON or OFF from rooms in the house (Balasubramanian, 2008). The remote control and monitoring of these appliances can be performed remotely via the Internet or GSM mobile phones. Furthermore, some machine can act smartly by reporting their problems to the service company. For example, the refrigerator might report a cooling problem to the maintenance company; this is much needed in case smart home inhibitors are in holiday.

Furthermore, by exploiting electronic tags in food's packages, clothes, dishes, etc smart home appliance can operate in intelligent ways. White goods companies such as Merloni Elettrodomestici, are introducing appliances that communicates with objects using RFID (Merloni, 2003). Washing machine can scan the load in it to adjust the wash cycle to be suitable for the fabrics used. The refrigerator also might warn the user about the expiry dates of some of the products inside it. In addition to that it can send automatically a shopping list to some home delivery services.

5.2 Safety and Security

Safety and security are important aspects of human life. Therefore, incorporating safety issues in smart home is an important requirement for most of the smart home occupants.
A number of products are available in the market that implements some of smart home concepts to deliver various aspects of safety, alarm and security. It is expected that the number of these products will grow in the nearby future. The general architecture of these systems consists of an appropriate set of interconnected sensors that monitors specific conditions or situations and communicates them to a local server which then transmits them to the concerned parties. These sensors could include smoke detectors, water leakage detectors, intruder detectors, power outage detectors, etc. In case of an alarm, both the home owner and the security company will be informed about the existence of the alarm. Having a smart system installed in the home, it will transmit detailed information specifying the exact location and the cause of the alarm. Furthermore, the system will allow the user to control some utilities of his home remotely. For example, in case that the home owner is expecting a home delivery and he cannot be in his house, the main gate can be opened to allow the postman to deliver the package. He can also close the doors and the main gate after the postman leaves the house. Via the Internet or his mobile phone, he could also switch on or off the heating/cooling system for a specific part of his home. In case the smart home occupants are in holiday, they could program the system to simulate the owner presence inside the home by switching on the home lights and switching them off at regular times.

5.3 Tele-health Care

The cost of providing health care particularly in developed countries continues to rise. This is attributed to many factors that include an increase in peoples' life expectancy coupled with a decline in birth rates which are resulting in an aging population. Add to this is the general increase in the salaries of health professionals and the cost of medication and diagnostic equipment. All these costs are putting a huge strain on the health budgets of the various government departments and agencies.
This means that the provisions for long-term health care within hospitals and clinics are severely restricted. However, those patients who may need long-term health care, such as elderly persons with heart problems, may not have people to look after them once they go home or they may simply prefer to live within their own normal environment for psychological, social or other similar reasons. The availability of a smart living environment, in the form of a smart home, will help alleviate some aspects of this problem and hence release resources that would otherwise go to long-term health support of an individual patient to other patients, enable early diagnosis of chronic health conditions, and make clinical visits more efficient due to the availability of objective information prior to the such visits (Botsis and Hartvigsen, 2008; Friedewald et al., 2005).
A number of studies have been reported in the literature that applies the smart home concept to deliver various aspects of telemedicine or tele-health (Liu et al., 2007). The general architecture of a system that delivers the required health-care services would consist

of an appropriate set of sensors that monitor specific medical conditions or situations and communicates it to a local server and then transmits it to the entity assigned to look after the patient. Many of the proposed telehealth system use multi-agent technology to arrive at an intelligent decision regarding the health state of the person being monitored (Rammal et al., 2008). The system described in (Tabar et al., 2006) uses a set of cameras to form a distributed vision processing setup. Analysis of the occupant's posture is then used to detect if the person suffered a fall. The assessment is based on a set of rules and once a fall situation is asserted an alarm gets sent to a call centre. A voice link can then be established between the occupant and the care centre.

5.4 Energy Metering

Environmental awareness is a major trend that has impacts on all businesses especially in the developed countries. The driving forces for this interest are rising energy costs and reducing greenhouse gas emissions. These factors are pushing to reduce energy usage and or optimize its usage in homes and commercial building through smart energy products.

In general, energy cost can be one fifth of building's operating costs, with lighting and heating or cooling using most of it in both residential and commercial buildings. Consequently, they are the main targets for energy reduction through smart energy usage. Utility companies are interested in smart ways to control the rising energy usage with minimum costs. They are reluctant to implement and deploy both load control and time-of-use (TOU) rates to help managing the load during peak demands. They would like to have the possibility, through load control, turn off some non-critical customer loads, such as HVAC, for short period of time, during peak demands. Also, they can use demand response mechanism where customers are informed about TOU rates. This will encourage the customers to manage their energy usage efficiently. These actions will help utility companies in reducing their peak demands and users to reduce their utility bills.

The use of smart electricity meters in smart homes will definitely be able to communicate with utility companies for the purpose of load control and demand response (Venables, 2007). Through Home Area Network (HAN), smart meters can communicate and control other appliances in the smart home, such as the HVAC system. In addition to that the smart home owner can take voluntary actions during demand/response periods to reduce his/her home consumption by turning off appliances, lights, etc. Having the household appliances interconnected will facilitate load control and demand response. For example, a dishwasher can be given the command to delay its start time to a time that has lower utility rate.

5.5 Environmental Control

Heating and cooling a house is the major post in the houses' utility bill, therefore, they deserve special attention when designing smart homes. Choosing good isolation materials for constructing homes that are suitable for their environment is the foundation to reduce these costs. By the deployment of smart sensors inside and outside the house, the houses' energy consumption can be managed efficiently. By collecting information from the distributed sensors inside the house such as thermostats, humidity, airflow, etc and from sensors outside the house such as thermostats, humidity, sunshine, wind speed meter etc, these information can be processed to control HVAC units and blinds/shades/rollers/windows. For example, in case that a window is opened for a pre-specified period

while the HVAC unit was on for that specific room, a decision can be made to stop that unit or to alarm user about this inefficient use of energy. In addition to that, an advice can be given to the user to shut the curtains in the night when the temperature outside is much lower than the one inside the house while the heating system was on.
The renewable sources of energy available to smart homes depend on their geographical locations. For example, in locations where the sun shines most of the year, it is advisable to have solar panels installed on its roof. In other locations where you have windy climate, it is advisable to have wind turbines installed in garden (Balasubramanian, 2008). By monitoring the environmental conditions inside and outside the house and the current home load condition, decisions can be made how to use the generated electricity. In case that the house is connected to smart electricity grid, the extra electricity can be fed back to the grid.

5.6 Information Access

The smart home should respond to the inhibitors needs and the change of these needs in unobtrusive and invisible way while remaining under the control of the users (Friedewald et al. 2005). By identifying the location of a person inside the house and by sensing his activity the environment can be adapted to suit his wishes. In case the person requires information it will be forwarded to the nearest display based on his location. Also his preferences and the accessed information should move with him when he changes his location. In case the smart home inhibitor was watching the news while he was shaving in the bath room, the same news channel will be displayed in the kitchen when he moved there to drink his coffee. In case the smart home inhibitor is reading a novel, the information related to novel could be retrieved from the Internet and displayed on the nearest display screen (Friedewald et al., 2005).

5.7 Smart Space for Kids

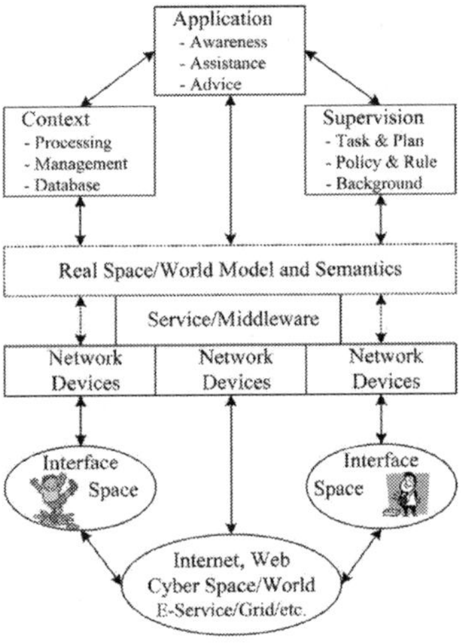

Fig. 8. System Architecture of UbicKids

The UbicKids system proposed by (Ma et al., 2005) aims at creating a smart hyperspace environment of ubiquitous care for kids. The system caters for both kids and parents. The main objectives of UbicKids are: (1) to develop a set of ubiquitous applications that help parents to take care of their kids through efficient, user friendly, reliable, secure and trustworthy services; (2) to study the social and psychological impact of ubiquitous kids care applications on the families; (3) to build a prototype smart hyperspace environment. The system architecture of UbicKids is illustrated in Fig. 8 (Ma et al., 2005).

6. Case Study

This section describes the architecture and design of a prototype smart home system that integrates a variety of wireless information and communication technologies with a variety of sensors and appliances. The prototype system enables an authorized user to monitor and control home devices equipped with a smart interface structure using a mobile phone. The system is embedded with authentication and verification mechanisms that provide secure end-to-end processing. The system is scalable through a flexible architecture that easily expands the portfolio of devices to be monitored and controlled. This system uses the mobile GSM/GPRS network for the external network. Where a mobile phone application using J2ME is designed to control and monitor various types of devices. It communicates with the home computer server which is connected wirelessly to the microcontrollers of the devices, being monitored or controlled, via a Bluetooth network. Although the Bluetooth technology used for the internal network has some limitations in terms of high cost and short range, and security concerns, its hardware size is small (Pooter, 2006). This technology can be replaced with the Zigbee technology since it is more suited for this type of application. Furthermore, it is cheaper, has better security and battery powered Zigbee devices require less power.

6.1 Smart Home Architecture

The architecture and the various technologies used in the design of the prototype system are described in Fig. 9. A comprehensive description of the system, its functionality, and services is given by (Al-Qutayri et al., 2010). The system consists of a Java-enabled mobile phone with a mobile application, a computer server with a Java application, a GSM modem connecting the server to the external network, a PIC microcontroller with a hardwired application connected to the devices, and Bluetooth adapters connecting the server and the microcontrollers to the internal Bluetooth network.

The system is connected to the external network using GSM network. A mobile phone communicates with the home server using Short Message Service (SMS). The mobile application which is written using Java 2 Micro Edition (J2ME) technology includes a friendly menu-driven Graphic User Interface (GUI). The GUI is used by the user to input all information to the system. It has a list to display the main system services, buttons to turn on, turn off or monitor a device, text fields to get the configuration information, and commands to move from menu to sub-menus. The system uses the Bluetooth P-to-P (Point-to-Point) protocol as the wireless internal network to connect the home devices' microcontrollers to the home server.

The home server is the core of the smart home system. It is connected to the GSM network via a GSM modem to send/receive SMS messages to/from the mobile phone. It also connects to home devices through the Bluetooth connection. Fig. 10 show the software architecture of the home server. The server application is written using the Java programming language.

Smartness is introduced to the devices in the prototype system by including a PIC microcontroller in each device in order to monitor and/or control it at the hardware level. It is also connected to a Bluetooth adapter to establish the Bluetooth P-to-P connection with the home server. An application program, which is written in C and downloaded to the PIC microcontroller, receives commands from the server, interprets them and sends the information to the devices. It also reads the status of devices and sends it to the server. The presence of the microcontrollers enables the introduction of smart control and monitoring functionality to the home devices. The level and sophistication of the smartness depends on the functions the device can perform.

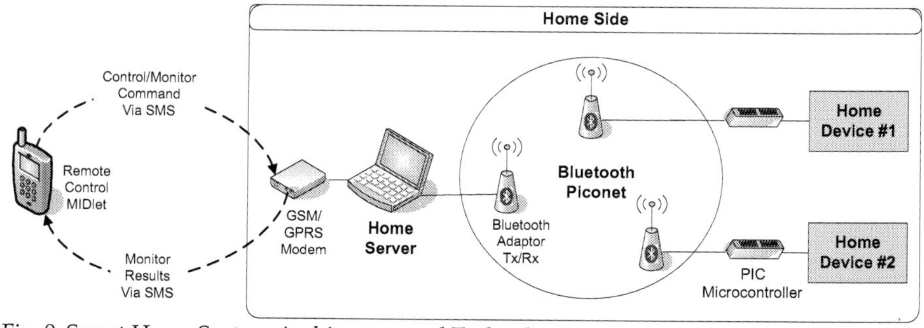

Fig. 9. Smart Home System Architecture and Technologies

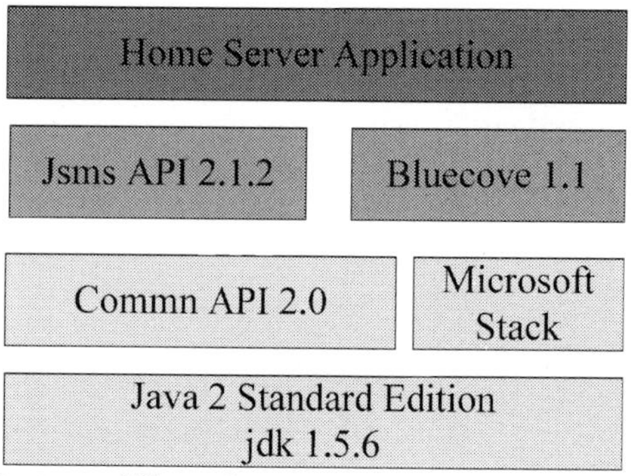

Fig. 10. Software Architecture of Home Server

6.2 System Services

The system provides three main services: monitor the settings of home devices and appliances, control the settings of home devices, and get status of devices periodically. Some additional minor services provided by the system are events and problems logging and notification of connections problems. Monitoring home devices consists of checking the setting of the devices. The monitoring command is sent via an SMS as illustrated in Fig. 11.

Controlling a home device includes changing the setting of the device by sending a command via an SMS containing the name of the device to be controlled and the control command as illustrated in Fig. 12. The control commands currently implemented are: turn off device, turn on device to a certain level, and change the level of device.
Instead of running the J2ME application in the mobile phone every time to get the device status, an SMS message is generated by the home server that periodically shows all devices status. The SMS message is sent according to a pre-programmed time interval. This time interval can be changed by the mobile phone application by sending a special command to the server.

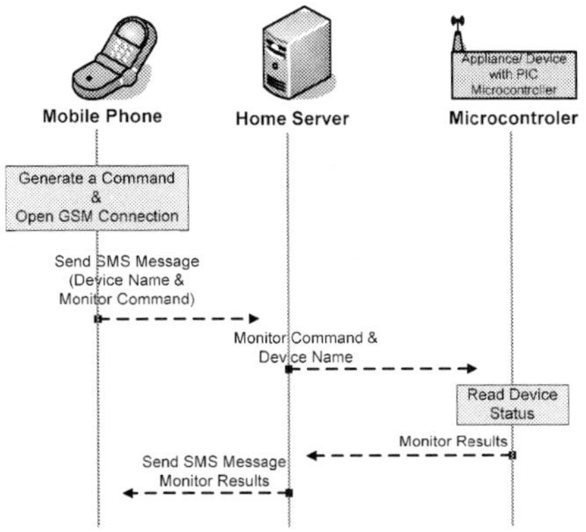

Fig. 11. Home Device Monitoring Process

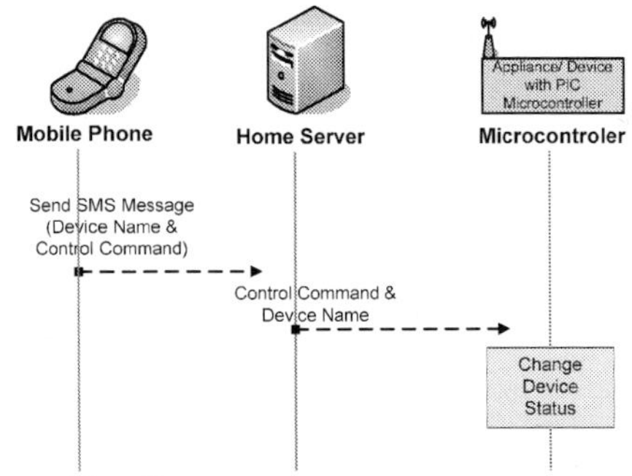

Fig. 12. Home Device Controlling Process

6.2 System Software Application Design

The system that enables monitoring and control of home appliances consists of three major software applications: mobile or hand-set, home server, and appliance controller.
The data flow diagram of the mobile application is depicted in Fig. 13. This application gets/sends data from/to two terminals. First time, the user configures the mobile application with two parameters: Home Server SIM card Number and Periodic time for SMS

notification. The Home Server connects to the GSM network using GSM modem and that has a SIM card Number.
The user controls the mobile application using the following set of commands:

- Set Level: select a level for a specific device
- Back and Select: enable the user to move from one category to another (Monitor a Device, Control a Device, and Configuration)
- Turn-on/Turn-off and Monitor: the command like buttons used by the user to send the command by pressing the button

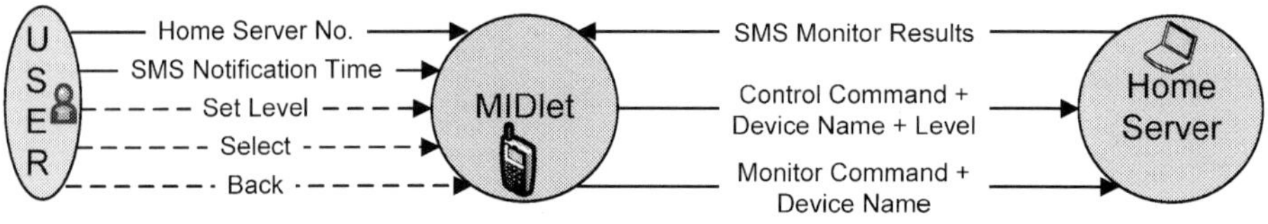

Fig. 13. Mobile Application Data Flow Diagram

The behaviour model of the mobile application is illustrated in Fig. 14. As can be seen in figure, the application moves or passes through four states to send a SMS message which contains the commands for home appliances.

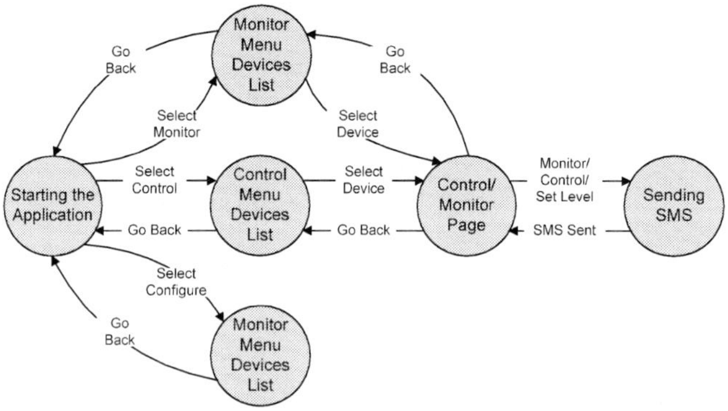

Fig. 14. Mobile Application Behavior Model

Fig. 15 show the data flow diagram model representing interaction between the home server and two appliances controlled by PIC microcontrollers. The home server communicates with each PIC microcontroller through a Bluetooth wireless connection. When the home server receives SMS message from a mobile phone, it extracts the message and checks its contents if the sender is an authorized one, otherwise it disregard the received message. The message could contain only three possibilities: Control a device, Monitor a device, or Set SMS notification time.

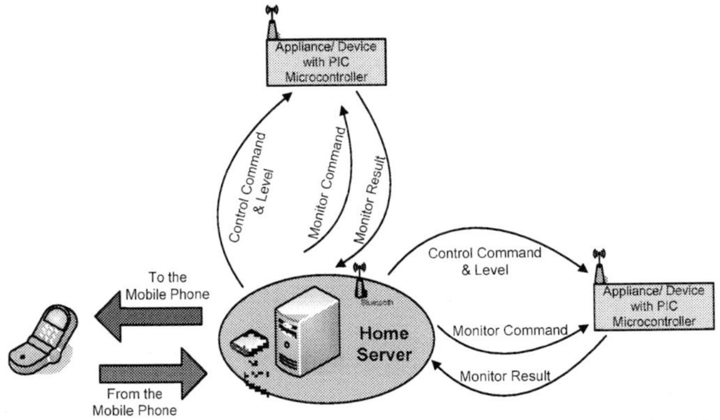

Fig. 15. Home Server Data Flow Model

If the message content is SMS notification time, the home server will set the new time. If the content is controlling a device the home server will look for the device. It assumes that each device at home has a smart interface, using a PIC microcontroller or another processor, and a Bluetooth MAC address. The server will open Bluetooth connection and then will send the control command (on/off) plus the level if the device has more than two control levels. After sending the command the connection will be closed by the server. In case of monitoring, the home server will open the Bluetooth connection and keeps it open until the PIC microcontroller replies with the device status. The SMS notification service can't be noticed from the data flow diagram therefore it is shown in the home server behavioural model in Fig. 16.

The behavioural model of the appliance controller application is shown in Fig. 17. The appliance is controlled via a PIC microcontroller. At the beginning, the PIC forces the Bluetooth adapter to change its mode to discoverable mode. In this mode the home server that acts as a master in the Bluetooth network scan and find the PIC Bluetooth adapter. The Bluetooth adapter accepts to open a connection with home server at the server request. The PIC microcontroller reads the sent command from the home server. For control command the PIC executes the command by setting the correct output at its output pins according to the sent command. For monitor command the PIC sends the status to home server.

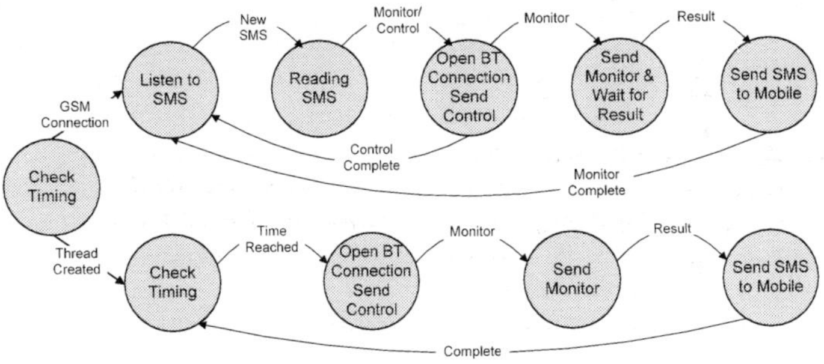

Fig. 16. Home Server Behavioural Model

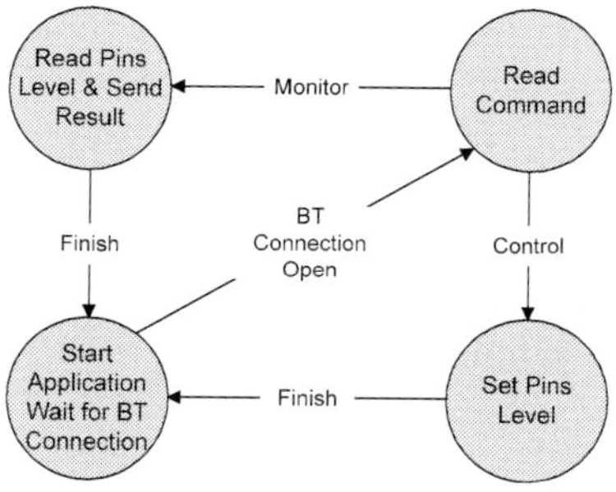

Fig. 17. Behavioural Model of Appliance Controller

6.2 System Security

Securing the complete system requires the incorporation of authentication as well as encryption layers. The authentication process is initiated through the GUI of the mobile phone once the application is enabled. The user is required to enter a username and a password. This information is then sent via SMS to the home server in order to establish the authenticity of the user through comparison with registered entries in the database. Once a valid user is identified the home server initiates a session that includes the phone number being used and a randomly generated number. The later number is used as an additional level of authentication and will expire once the session ends.

In this system, users are assigned access levels that define the actions they are allowed to perform. Security of the communication between the mobile station and the GSM/GPRS modem connected to the home server is achieved by the GSM encryption standard implemented at both ends. This is a highly robust method and hence additional security is not needed.

The most vulnerable part of the system is the connection between the home server and the various devices in the home. This is due to the use of Bluetooth to achieve connectivity and hence device monitoring and control. Bluetooth has inherent security problems that will expose the system to hacking should they be not addressed (Ferro and Potorti, 2005). Therefore, securing Bluetooth P-to-P communication required the implementation of an encryption/decryption process between the home server and the smart home devices that use Bluetooth wireless technology. This was implemented using a simplified form of AES (advanced encryption standard) algorithm. AES is a well known robust encryption standard that is widely used in various applications. This block cipher is relatively easy to implement, and does not require large amount of memory (Daemen and Rijmen, 2002).

6.3 Complete System Results

The system functionality and services were tested through verification and validation at all levels. This process included testing the user login and authentication, and remote monitoring and controlling devices/appliances at the home. The security features implemented to secure the communication between the home server and the remote devices were also tested in the process. The prototype system was evaluated by connecting the home server to a two-level device and a multiple-level device. The following paragraphs demonstrate the security aspects discussed earlier and the use of a three-speed DC Fan as an example of a multiple-level device to be monitored and controlled.

As soon as the user starts the application on the mobile hand-set the MIDlet main login page appears as shown in Fig. 18(a). Under the Menu button page shown in Fig. 18(b) the user can select to either Register or Login. Selecting the Register option enables the user to register new users and add them to the database with the appropriate privilege level as shown in Fig. 18(c) below. If the user chooses the Login command the username and password get added to the URL and the http request is initiated to access the web server. An alert appears to notify the user that it is going to start an http request then a waiting page will appear. Depending on the user privilege that is stored in the database the appropriate list of commands appear. Should the login not be successful the process will go back to the initial login page.

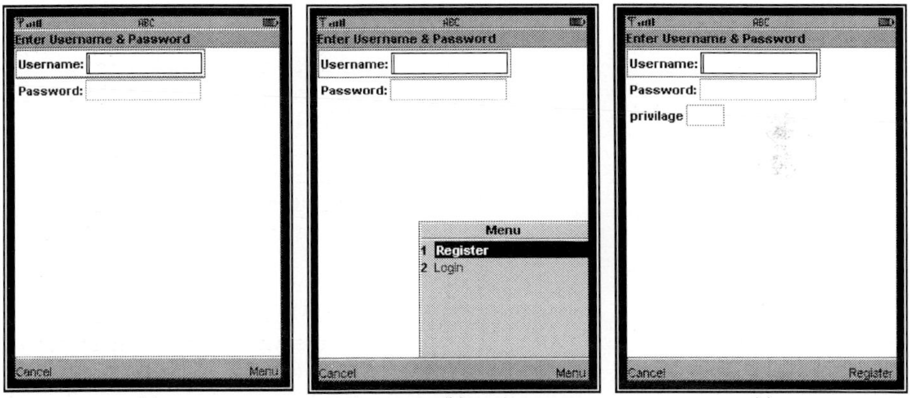

(a) (b) (c)

Fig. 18. (a) Main Page (b) List of Commands (c) Registration Page

The home device monitoring and control is illustrated in Fig. 19. To control the Fan the user chooses control a device from the main menu of the mobile application as shown in Fig. 19(a). The main menu also includes: monitor a device to check the status of a device and configuration to set the time interval for periodic notification of all devices' status. When the user selects the control mode, all connected home devices will be listed as shown in Fig. 19(b). Selecting a device will result in displaying a control page which has two buttons as shown in Fig. 19(c). Pressing Turn-Off will result in sending an SMS message to the home Server to turn-off the selected device. Clicking on Turn-On opens a new page as shown in Fig. 20(a). If the device to be controlled is a two-level (ON or OFF) device, this page will not be displayed. The next step is to select one of the three speeds for the Fan. An SMS is then automatically sent to the home server. The SMS includes the Fan name, the turn-On control command and the level information (speed two in Fig. 20). The home server would then open a Bluetooth connection with the PIC microcontroller of the Fan device on a certain MAC address. When the Bluetooth adapter on the device side accepts the connection, the home server sends setting the speed command. The Fan receives the command and changes its state to the appropriate speed.

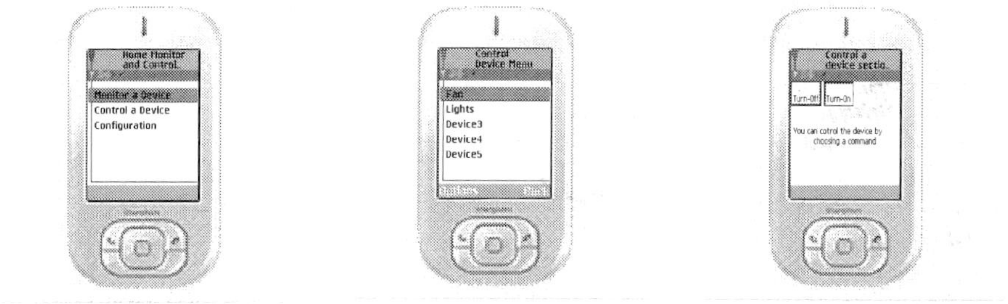

(a) (b) (c)

Fig. 19. (a) Main Menu (b) Control Menu (c) Control Page

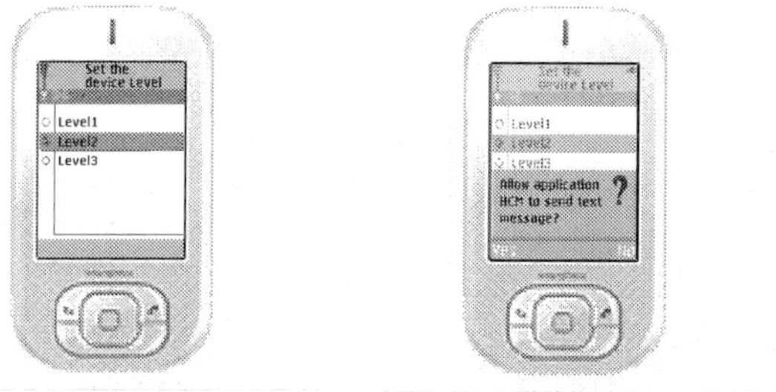

(a) (b)

Fig. 20. Set the Speed for Fan (a) Set Level (b) Send Control Message

In this study the impact of overall system delay and failure of SMS delivery due to various communication and processing aspects was not tested directly. The rational for not implementing that is that the system is a prototype one and through assessment of its reliability and hardening of its security are not necessary. Having said this, the system has a regular update mode that can be used to check the operating status of the various appliances. This is not envisaged as a replacement for a proper mechanism to take care of delays and failure issues.

7. Conclusions

This chapter presented some of the state of the art technologies and associated applications in the field of smart homes. It gave an overview of the major wireless communication technologies that form a fundamental part of the infrastructure of modern smart homes. Some of those technologies are integrated within sensing and networking devices such as Zigbee, Bluetooth, RFID, and WiFi. Other wireless technologies, such as the GSM, are more of a wider format that can form large network and yet can integrate with the other ones dedicated for short range. The paper also briefly discussed some of the modern sensors that can be used in smart homes. Many of them are of the embedded ubiquitous type that is equipped with wireless communication capabilities and can connect to other devices.

The application areas discussed include appliances monitoring and control, safety and security, telehealth care, energy saving, environmental control, and information access.

Some of these areas are more developed than other, however, all those areas are either already available or they are excepted to be deployed in the near future. The chapter included a case study of a complete end-to-end smart home system that is used to monitor and control home appliances using a mobile phone. The prototype system used GSM as the external network and Bluetooth as an internal network. However, other suitable wireless technologies can be used on the same architecture. The system enables two way control and has automatic updating service that informs the user about the status of the devices at regular intervals.

Given the advanced status of the constituents of smart homes, it is expected that many of the existing homes will be turned smart in the not too distant future. So, the future vision of the smart home is getting closer, but the designers need to spend more time to learn how people live within the bounds of their homes.

8. References

Al-Qutayri, M., Barada, H., Al-Mehairi, S., and Nuaimi, J. (2008) "A Framework for an End-to-End Secure Wireless Smart Home System," IEEE Systems Conf., pp. 1-7

Al-Qutayri, M., Barada, H., and Al-Mehairi, S. (2010) "Integrated Secure Wireless System for Smart Home Monitoring and Control," Int. Journal of Computer Aided Engineering and Technology, Vol. 2, Nos. 2/3, pp. 181-198

Al-Qutayri, M.A. and Jeedella, J.S. (2010) "Smart Homes: Technologies and Challenges," Int. Journal of Computer Aided Engineering and Technology, Vol. 2, Nos. 2/3, pp. 125-144.

Augusto, J. (2007) "Ambient Intelligence: the Confluence of Ubiquitous/Pervasive Computing and Artificial Intelligence," In Intelligent Computing Everywhere, Springer, pp. 213-234

Augusto, J. C. and Nugent, C. D. (2006) "Smart Homes Can Be Smarter," In Designing Smart Homes - The Role of Artificial Intelligence, pp. 1–15.

Balasubramanian, K. and Cellatoglu, A. (2008) "Improvements in Home Automation Strategies for Designing Apparatus for Efficient Smart Home," IEEE Transactions on Consumer Electronics, Vol. 54, No. 4, pp. 1681-1687.

Baldauf, M., Dustdar, S., and Rosenberg, F. (2007) "A Survey on Context-Aware Systems," Int. Journal of Ad Hoc and Ubiquitous Computing, vol. 2, no. 4, pp. 263-277

Botsis, T. and Hartvigsen, G. (2008) "Current Status and Future Perspectives in Telecare for Elderly People Suffering from Chronic Diseases," Journal of Telemedicine and Telecare, Vol. 14, No. 4, pp. 195-203

Brand, A. and Aghvami, H. (2002) Multiple Access Protocols for Mobile Communications: GPRS, UMTS and Beyond, Wiley

Bricon-Souf, N. and Newman, C. R. (2007) "Context awareness in health care: A review," Int. Journal of Medical Informatics, vol.76, no.1, pp.2-12

Carettoni, L., Merloni, C., and Zanero, S. (2007) "Studying Bluetooth Malware Propagation: The BlueBag Project," IEEE Security & Privacy, vol.5, no. 2, pp.17-25

Chan, H. and Perrig, A., (2003) "Security and Privacy in Sensor Networks," IEEE Computer, Vol. 36, Issue 10, pp. 99-101.

Choi, J., Shin, D., and Shin, D. (2005) "Research and Implementation of the Context-Aware Middleware for Controlling Home Appliances," IEEE Trans. on Consumer Electronics, vol. 51, no. 1, pp. 301-306

Cook, D. J. (2006) "Health Monitoring and Assistance to Support Aging in Place," Journal of Universal Computer Science, vol. 12, no. 1, 2006, pp. 15-29

Cook, D.J., Youngblood, M., Heierman, E., Gopalratnam, K., Rao, S., Litvin, A., and Khawaja, F. (2003) "MavHome: An agent-based smart home," First IEEE Int. Conf. on Pervasive Computing and Communications (PerCom'03), pp. 521–524.

Daemen, J. And Rijmen, V. (2002) The Design of Rijndael: AES – The Advanced Encryption Standard (Information Security and Cryptography), Springer

Dargie, W. (2009) Context-Aware Computing and Self-Managing Systems, Chapman & Hall

Darianian, M. and M. P. Michael (2008) "Smart Home Mobile RFID-Based Internet-of-Things Systems and Services" Int. Conf. on Advanced Computer Theory and Engineering (ICACTE08), pp. 116-120

Decker, C., Krohn, A., Beigl, M., and Zimmer, T. (2005) "The Particle Computer System," Int. Symp. on Information Processing in Sensor Networks, pp. 443-448.

Dengler, S., Awad, A., and Dressler, F. (2007) "Sensor/Actuator Networks in Smart Homes for Supporting Elderly and Handicapped People," Int. Conf. on Advanced Information Networking and Applications Workshops, pp. 863-868

Estrin, D. Culler, D. Pister, K. and Sukhatme, G. (2002) "Connecting the physical world with pervasive networks," IEEE Pervasive Computing, Vol.1, Issue 1, pp.59-69.

Ferro, E. And Potorti, F. (2005) "Bluetooth and Wi-Fi wireless protocols: a survey and a comparison," *IEEE* Wireless Communications, Vol. 12, Issue 1, pp. 12-26

Friedewald, M., Da Costa, O., Punie, Y., Alahuhta, P., and Heinonen, S. (2005) "Perspectives of ambient intelligence in the home environment," Telematics and Informatics Vol. 22, pp. 221-238.

Garlan, D., Siewiorek, P.D., Smailagic, A., and Steenkiste, P. (2002) "Project Aura: Toward Distraction-Free Pervasive Computing," IEEE Pervasive Computing, Vol. 1, No. 2, pp. 22–31.

Gislason, D. (2008). Zigbee Wireless Networking, Newnes

Haryanto, R. (2005) Context Awareness in Smart Homes to Support Independent Living, MSc Thesis, University of Technology, Sydney, Australia, 2005

Helal, S., Mann, W., El-Zabadani, H., King, J., Kaddoura, Y., and Jansen, E. (2005) "The Gator Tech Smart House: A Programmable Pervasive Space," IEEE Computer Magazine, no. 3, pp. 64-74

Ilarri, S., Mena, E., and Illarramendi, A. (2008) "Using Cooperative Mobile Agents to Monitor Distributed and Dynamic Environments," Information Sciences, vol. 178, no. 9, pp. 2105-2127

Juels, A. (2006) "RFID Security and Privacy: A Research Survey," IEEE Journal on Selected Areas in Communications Vol. 24, Issue 2, pp. 381-394.

Kientz, J., Patel, S., Jones, B., Price, E., Mynatt, E., and Abowd, G. (2008) "The Georgia Tech Aware Home," ACM Conference on Human Factors in Computing Systems, pp. 3675-3680

Labiod, H., Afifi, H. and De Santis, C. (2007), WI-Fi, Bluetooth, Zigbee and Wimax, Springer.

Leopold, M., Dydensborg, M., and Bonnet, P. (2003) "Bluetooth and Sensor Networks: A Reality Check," Int. Conf. on Embedded Networked Sensor Systems, pp.103-113

Liu, Y., Bacon, J., and Wilson-Hinds, R. (2007) "On Smart-Care Services: Studies of Visually Impaired Users in Living Contexts" Int. Conf. on Digital Society (ICDS), 2007

Ma, J., Yang, L., Apduhant, B., Huang, R., Barolli, L., and Takizawa, M. (2005) "Towards a Smart World and Ubiquitous Intelligence: A Walkthrough from Smart Things to Smart Hyperspaces and UbicKids," Journal of Pervasive Computing and Communications, Vol.1, No.1, pp. 53-68

Mann, W., El-Zabadani, H., King, J., Kaddoura, Y., and Jansen, E. (2005) "The Gator Tech Smart House: A Programmable Pervasive Space," IEEE Computer, Vol. 38, Issue 3, pp. 64-74

Marsa-Maestre, I., Lopez-Carmona, M., Velasco, J., and Navarro, A. (2008) "Mobile Agents for Service Personalization in Smart Environments," Journal of Networks, Vol. 3, No. 5, pp. 30-41.

Merloni (2003). "Merloni Unviels RFID Appliances," RFID Journal

Mozer, M. C. (2005) "Lessons from an Adaptive House, " In D. Cook & R. Das (Eds.), Smart environments: Technologies, protocols, and applications, (pp. 273-294), Wiley & Sons

Oh, Y., and Woo, W. (2004), "A Unified Application Service Model for ubiHome by Exploiting Intelligent Context-Awareness," Int. Symp. Ubiquitous Computing Systems, LNCS, pp. 192-202

Oh, Y. and Woo, W. (2005) "A Unified Application Service Model for ubiHome by Exploiting Intelligent Context-Awareness," In Ubiquitous Computing Systems, Springer, pp. 192-202

Pooter, B. (2006) "Bluetooth Security Moves," Network Security, Vol. 2006, Issue 3, pp.19-20.

Rammal, A., Trouilhet, S., Singer, N., and Pécatte, J.M. (2008) "An Adaptive System for Home Monitoring Using a Multiagent Classification of Patterns," Int.l Journal of Telemedicine and Applications, Vol. 2008, Article ID 136054

Roduner, C., Langheinrich, M., Floerkemeier, C., and Schwarzentrub, B. (2007) "Operating Appliances with Mobile Phones - Strengths and Limits of a Universal Interaction Device," in Pervasive Computing, LNCS, Springer, pp. 198-215

Satyanarayanan, M. (2001). "Pervasive Computing: Vision and Challenges," IEEE Personal Communications," vol. 6, no. 8, pp. 10–17

Schaefer, R., Mueller, W., and Groppe, J. (2006) "Profile Processing and Evolution for Smart Environments," Proceedings of 3rd Int. Conf. on Ubiquitous Intelligence and Computing (UIC-06), LNC, pp. 746-755

Tabar, A. M., Keshavarz, A., and Aghajan, H. (2006) "Smart Home Care Network using Sensor Fusion and Distributed Vision-based Reasoning," Proceedings of the 4th ACM Int. workshop on Video surveillance and sensor networks, pp. 145-154.

Thompson, T. J., Kumar, C. B., and Kline, P. J. (2008), Bluetooth Application Programming with the Java APIs, Morgan Kaufmann

Velasco, J., Marsá-Maestre, I., Navarro, A., López, M., Vicente, A., Hoz, E., Paricio, A., and Machuca, M. (2005) "Location aware services and interfaces in smart homes using multiagent systems," Proc. Int. Conference on Pervasive Systems and Computing (PSC'05), USA

Venables, M. (2007). "Smart Meters Make Smart Consumers [Analysis]." Engineering & Technology 2(4): 23-23.

Weiser, M. (1991) "The Computer for the 21st Century", Scientific American, vol. 165, pp. 94-104

Want, R. (2006). "An Introduction to RFID Technology," IEEE Pervasive Computing, Vol. 5, Issue 1, pp.25-33.

Yamazaki, T., (2007) "The Ubiquitous Home," Int. Journal of Smart Homes, Vol. 1, No. 1, pp. 17-22

Zambonelli, F., Jennings, N., Omicini, A., and Wooldridge, M. (2000) "Agent Oriented Software Engineering for Internet Applications," In Coordination of Internet Agents: Models, Technologies and Applications, Springer

3

Selected Home Automation and Home Security Realizations: An Improved Architecture

K. Balasubramanian and A. Cellatoglu
European University of Lefke
North Cyprus, Turkey

1. Abstract

The main objective of this presentation is to give the design of main equipments for intelligent home meeting the modern requirements and satisfying most living standards of consumers. In this endeavor the home automation considerations of this presentation focus on manual and remote control of selected appliances, timed setting of switching the appliances and personal digital home assistant software that brings the attention of the resident about the tasks of the day to be performed. Home security concerns of the system are the incorporation of *i.* real time audio visual system that permits regulated admittance of the visitors after approval from the resident and *ii.* remote alerting the resident upon detecting the fire or intruder. The design approach is based on the support of the central web server and monitoring unit and is meant for medium sized residential complex constituting many flats. In order to self support the energy needs of the flats to an extent a cost effective dual energy extraction unit generating electricity from the renewable energy resources is also included in the system and fixed in each flat. The performance of all schemes presented here are compared and analysed for their adaptation to any installation. The hardware devices and components used are commonly available in practice and the realization of the system for any further expanded requirements would be quite feasible and easy.

2. Introduction

Home automation activities are becoming increasingly important nowadays in providing more comfort and security for the home residents. Reports are available in the past concerning the development of devices and units needed for implementing the smart home (two websites, 2009; Jorge Caleira Nunes et al, 2004; Renato Nunes, 2003 and Balasubramanian and Cellatoglu, 2008). Each implementation deals certain aspects of automation satisfying partial requirements of the consumers. This project deals with the design of home automation apparatus satisfying essential requirements of automation needed for comfortable stay and pleasant living in a flat of multi storied building. Also, generating electrical energy from natural resources and their utilization schemes for feeding the apparatus are implemented in the system as to promote contributing to alternate energy resources and to reduce the cost constraints of energy consumption. These schemes are designed to extract maximum solar and wind energy and used to feed selected appliances of the residential flat. In the absence of sun and wind energy the power system lines supplying electricity to the flat would take care of powering the selected appliances. Furthermore, as an additional security concern, an intruder detection system is installed in the system which when detects an intruder would dial automatically a sequence of digits programmed in the system as to give remote intimation for the intrusion found. As internet and telephone communication are quite popularly used nowadays, remote control schemes of selected appliances in flat are also included in the system as to serve the day to day urgent needs and also for security concerns. This scheme facilitates the control of appliances distributed in the flat by operating from any other room. Furthermore, a Home Assistant software installed in the system refers the home data base every morning and brings the list of activities to be performed on that day to screen as to alert the user to be ready for solving the issues of the day.

The software cannot be accessed by any unknown person due to password requirement. Necessary firewall is incorporated in the web server as to avoid further interruptions due to unauthorized interruption and to block viruses.

3. Selected Remote Control Techniques

The control techniques which are most viable and for easy implementation to home automation system are presented here.

3.1 Web Based Control

Internet usage has become a common means of sharing and exchanging information between users. By activating web page setup for remote control purposes we can selectively issue commands to switch ON or OFF the selected appliances in home. This is an active method of controlling the appliances wherein command can be issued after knowing the status of the appliance.

3.1.1 Hardware Requirements of Web Based Interactive Control

Fig.1 shows simple schematic of the configuration of the internet based remote control activity. A Relay Board carrying register IC (Integrated Circuit) and an array of relays is the important final control unit of the home automation system. The control word in the relay register commands to switch ON/OFF of the appliances. The host PC is connected to web server through LAN which extends internet facilities to PCs of other flats as well. The server in turn is connected to the relay board which controls the switching of appliances. A special I/O card having units to access 16-b command word from an input port and also to send a 16-b status word to output port is extended to the server. Thus, the command word received from web is driven to the relay board and the status word read from a logic circuit is given back to the web server.

3.1.2 Software Requirements of Web Based Control

The website for remote control is designed with HTML. The password issues are resolved with Java Script. Once the website is open in a remote PC by the user a control table appears in the monitor screen providing options for the user for commanding the remote switching. The model of the control table that would appear in the screen is given in Table 1. All active items are programmed with ASP (Active Server Page) so as to communicate between the internet accessed PC and the web server kept in the home buildings. Visual Basic program (VB) is used in the server to communicate with the website and also to the relay board.

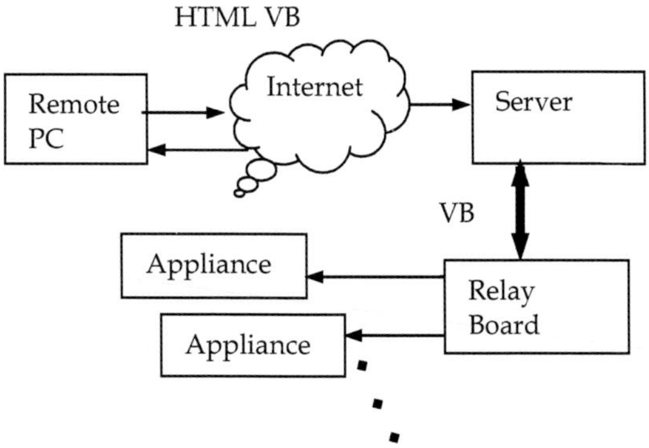

Fig. 1. Simple Schematic of Web Control

Item	Appliance	Action	Action
1	Fridge	ON	OFF
2	Air-Conditioner	ON	OFF
3	Lamp-1	ON	OFF
4	Lamp-2	ON	OFF
5	Fan-1	ON	OFF
6	Fan-2	ON	OFF
7	TV	ON	OFF
8	Home Theater	ON	OFF
9	Washing Machine	ON	OFF
10	Water Pump	ON	OFF
11	Appliance-1	ON	OFF
12	Appliance-2	ON	OFF
13	Appliance-3	ON	OFF
14	Appliance-4	ON	OFF
15	Status	Yes	No
16	Quit/Submit	Submit	No

Table 1.Control Table Appearing in the Monitor

3.1.3 Sequence of Operations

The sequence of operations performed for remote control actions are as follows. The resident at remote location opens the website and accesses the control table after giving user name and password. If he wants to know the present switched status of appliances before giving the command he clicks the *status* icon (item 15) in the control table. This action causes the message from this web page to reach the server. Consequently, in the web server the status word from the relay register in the relay board is sensed and returned back to the remote PC and placed in the 16-b status register. This 16-b status register has two most significant bits reserved for fire flag and intruder flag and the rest 14 bits denote the status of ON or OFF of the 14 appliances considered in the system. The appearance of the bits as '1's or '0's indicate the appliances switched ON or OFF respectively. Therefore looking at the status register the user can decide to take further steps. For instance, if he sees the Fan-1 (item 5) in switched ON condition and if he wants to switch OFF it he gives OFF command for Fan-1 by clicking OFF field of item No. 5. This flag can be checked immediately in the corresponding position of the command register. This action reaches the server and to the relay register in the relay board causing the Fan-1 to switch OFF and returns back to the status register in control table. This can be checked with 16-b status register.

If the user wants to control any further appliance he clicks the desired field and continues the operation. The command and status registers will be updated after issuing each command. In the end *quit* option with *submit* is given to finish the current sequence of operation.

At the instant of accessing the control table if the fire is detected, then the first *msb* would be '1'. On the other hand if intruder is detected then the second *msb* would be '1'. This is additional information, besides the alerting action given, for fire or intruder occurrence happened in the past 6 hours. This period is programmable by adjusting the period of a timer implemented with a monostable multivibrator linked to the fire detector and that with the intruder detector.

3.1.4 Residential Status Influencing the Control

If the resident is inside home then remote control becomes void and local manual control of appliances becomes effective. In order to ascertain the status of this a sense switch 'S' linked to the door in the home is activated to give a bit as '1' for inside and '0' for outside. Each of the 14 selected appliances is controlled by its relay commanded either by manual control 'M' or remote control 'R' depending upon the sense switch 'S'. Table 2 gives the truth table for designing a logic circuit 'L1' that produces a bit 'Y' for controlling the relay circuit of the concerned appliance. The Boolean expression for Y is obtained as

$$Y = S'.R + S.M \tag{1}$$

The logic circuit L1 that drives the bit *Y* for the relay board is shown in Fig. 2.

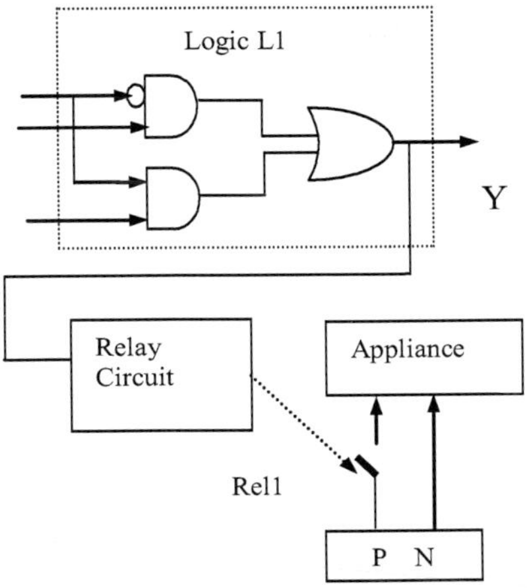

Fig. 2. Logic L1 Extending Control to Appliance

S	M	R	Act	Y Out
0	0	0	R	0
0	0	1	R	1
0	1	0	R	0
0	1	1	R	1
1	0	0	M	0
1	0	1	M	0
1	1	0	M	1
1	1	1	M	1

Table 2. Truth Table for Logic L1

3.2 Email Based Control

Email is another commonly and frequently used message transfer communication media and this is attempted to implement the remote control of appliances.

3.2.1 Hardware Requirements of Email Based Control

Email based control for commanding the relay board requires the relay board to be extended to the home PC. Although any number of appliances can be controlled we again assume the same 14 number of appliances to be controlled by email. Executable file (**.exe*) is stored in the hard disc of the home PC and the instructions in this **.exe* file issues commands to the relay board. The PC has internet connection and is able to receive emails as well. Those emails which are set for the purpose of controlling the appliances are identified and commands are given accordingly.

3.2.1 Software Involvement

The email platform has to be moderated to prepare the home PC ready for email based control.

3.2.1.1 Sequence of Operations Involved in Preparing the PC Microsoft Outlook Platform

Rules are created with Microsoft Outlook as to suit the custom requirements.

3.2.1.2 Control Action by Email Control

When once the email account is set up the system is ready to receive the email to the concerned email-id abc@mail.com. This email should bear the subject *FanON*. When once the mail with subject *FanON* arrives the in-box, instantly the *FANON.exe* file stored in the memory is executed. As a consequence, instructions in this *FANON.exe* file would send a control word to the relay board as to switch ON the fan.

- Open *Remote Commands* folder

  Go to Tools menu and select *Rules and Alert* option
- *Rules and Alert* page opens and Choose *New Rule* option and set changes.
- *Rules Wizard* page appears now and two steps are seen in this page.

  In Step 1 *select a Template* and opt the desired.

  In Step 2 Select user account by giving email-id and select folder for saving commands. Example: abc@mail.com

  move it to *Remote Commands* folder when this page is finished, again
- *Rules Wizard* page appears with selected options.

In this activity provide the specific word.

Step 1: Specific word in the subject

Step 2: From abc@mail.com and Fan*ON* and click *next*

to go back to

- *Rules Wizard* page
- Step 1: *Start Application* : *run a script*

  Step 2: From abc@mail.com and FanON on this machine only

  Start *FANON.exe* move it to *Remote Commands* folder.

  and complete this page to show
- *Rules and Alert* page

  OK to complete setting platform

Fig. 3. Preparation of Base for Email Control

3.3 SMS based Control

Sending text messages to a cellular phone from other cellular phone has also become a common practice nowadays. This method of communication depends on the cellular telecommunication network. We now use this facility for controlling the remote home appliances by SMS sent from a cellular phone.

3.3.1 Hardware and Setup requirements of SMS based Control

Several cell phones have connection ports to enable them to interface with PC or other equipments (Mobile telephone directory, 2009) for transferring files from phone to its peripherals. Using this feature we now use a decoder to the port for converting the digit received from SMS as a drive code for the register in the relay board. Whatever may be communication medium of issuing commands to the appliances the final control elements are the relays stacked in relay board. Fig.4 shows a simple schematic of the SMS based control system.

The user sends SMS message to control an appliance in the flat from another cell phone. For instance he wants to switch ON Fan-1, as indicated in the control table of Table 1. As the code for Fan-1 being '5' he sends an SMS from a cell of known number *123456* {say} to the home cell phone just the code '5 ON'. This digit as the text message reaches the cell phone and is saved in in-box. The sending mobile number *123456* also is saved. A software adjustment is made in the mobile system phone that if the message arrives from the mobile of No. *123456* then after identifying the digit received the digit '5' (0101) is sent together with '1', for ON command, to the output port. This makes the ON/OFF' flag be '1' for ON action to be commanded. The decoder decodes the command 5 (0101) into {0000000000010000} and provides the drive control bit '1' for {ON/OFF'} selecting the appliance in 5th position for ON action. Consequently the relay circuit of Fan-1 would be switched ON irrespective of what the status it was before.

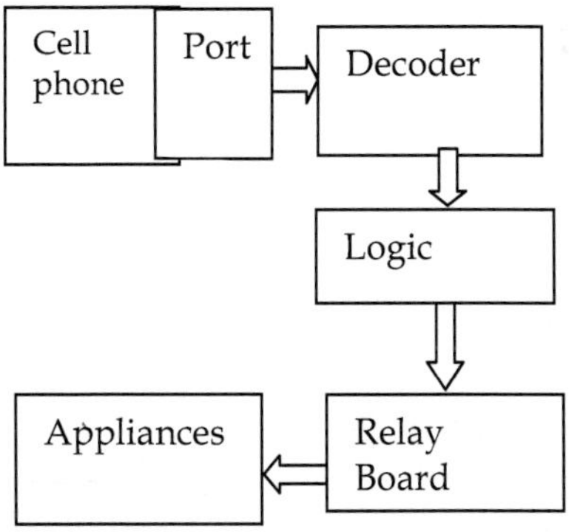

Fig. 4. Simple Schematic of SMS Based Control

3.3.2 Relay Register Circuit

The relay register needs to retain the flags (control bits) of the other appliances while changing the flag of the present command. This is realized by the circuit shown in Fig. 5.

The JK Flip flops working as flag bits of the relay register has asynchronous bits Preset (*PR'*) and Clear (*CLR'*) and the command is exerted through them. The figure shows only flip flop. The Enable bit (En) arrives from the decoded appliance code and the *ON/OFF'* bit arrives due to action code. For instance if the 5th appliance is commanded then *En* bit of the 5th flip flop becomes '1' and if ON command arrives *ON/OFF'* bit becomes '1' causing *PR'* to '0' as to set the flip flop ('1'). Obviously if *ON/OFF'* is '0' then flip flop would be cleared.

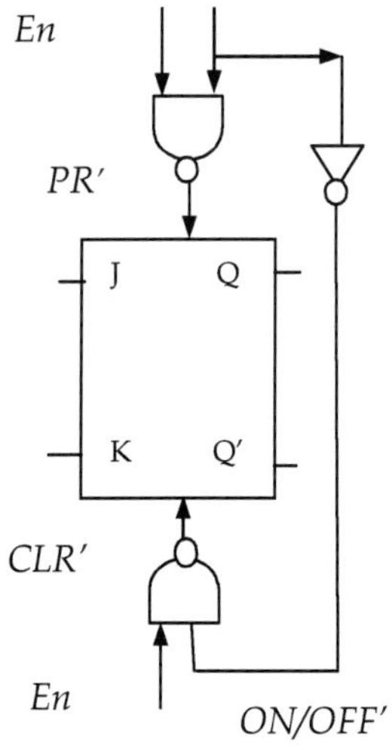

Fig. 5. Updating the bit of Relay Register

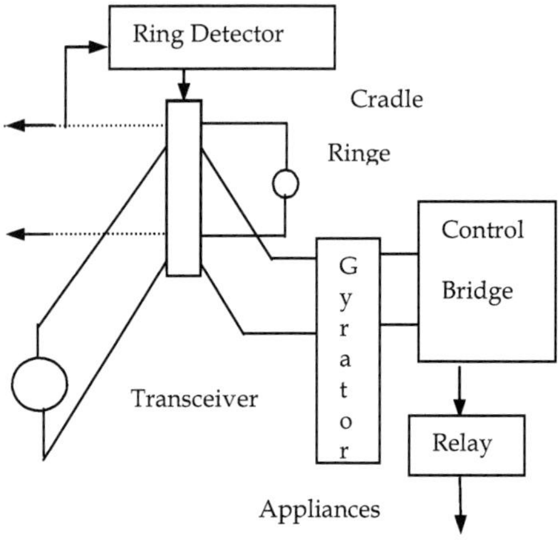

Fig. 6. Telephonic Control of appliances

3.3 Telephonic Method of Control

Landline telephony has been serving as the major communication media used for tele-conversation. Although the launching of cellular telephony and the internet based voice communication has reduced its usage, landline telephonic network still enjoys its service due to its simplicity and cost effectiveness offered to subscribers. Using landline telephone controls were issued (Balasubramanian, 2003) by dialing additional digits for switching ON or OFF of the electrical appliances in home. An extension card attached to the home telephone identifies the digit, decodes it and controls switching the appliances. Its principle is outlined in Fig.6.

In order to control switching the appliances in home the user dials the digits of home telephone. This gives ring in home telephone. The extension card attached to it counts the ring cycles and actuates a relay to close the cradle contacts despite the handset unmoved from cradle. The closing of contacts extends a gyrator and control bridge across the lines simulating the condition of establishing connection to home set. The additional digit dialed is decoded and the corresponding appliance is switched ON/OFF by the relay circuit.

3.4 Comparison of Remote Control Techniques

In order to compare the features of remote control techniques considered here a simple hardware scheme employing these control methods in home automation unit is shown in Fig.7. It does not, however, show the intruder detector and fire detector alarm units. Assuming all selected control schemes are available in the flat their features and performances are compared. The final control element of appliances of all control units are delivered from the relay board as commanded by word in relay register. At any instant of issuing the command from remote location the concerned control unit will write the control word into the relay register through the logic circuit and consequently the selected appliance would be operated. If the resident is in home as identified by the status switch 'S' the manual control bit would replace the remote control bit in the relay register.

3.4.1 Web Method

This is active method of controlling the appliances. The user can learn the status of the appliance whether it is switched ON or OFF and then issue order accordingly. The hardware required for this unit is the microcontroller interfaced to the web server. The command word is written into the relay register from the web through microcontroller and logic circuit. Access for control is almost instantaneous so long as the internet is of high speed and is active. Drawbacks of this method are the interruptions arising in internet at web server level or in channels including the satellite media.

3.4.2 Email Method

Email method of control is a passive control type that it does not check the present status of the appliance whether switched ON or OFF and simply issues order to selected appliance to either switch ON or OFF. While web control technique is almost instantaneous in executing the orders while the internet is active, the email method takes relatively longer time depending upon the type of the email platform employed. It also needs internet connection in order to carry on control activities. Since email is given from a remote PC web mail feature could be employed for controlling appliances.

3.4.3 SMS Control

Since cellular network is independent from internet and landline telephone network it needs just cellular phone for passing the control digits to the relay register. Neither PC nor its connection to internet is needed for control through SMS. From remote mobile phone an SMS need be sent to the home mobile phone for effecting the control of the appliance. The hardware requirement is therefore simpler than other methods of control.

3.4.4 Telephone Dialling

Landline telephone dialing is another simple method of control. Here the landline telephone is attached with an extension card and kept in home such that when the control digit is dialed from external phone it is identified and passed to the relay register for effecting the control. This control unit is simple and is involved in telephone loop independent of the internet connectivity. Unlike web based control there is no facility to check the status of the appliance before commanding the control.

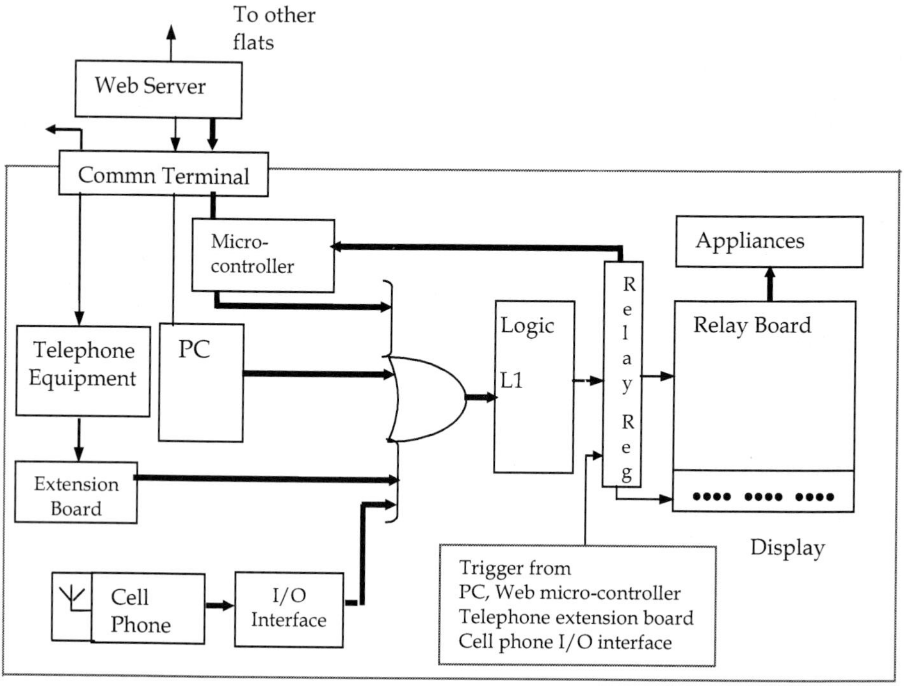

Fig. 7. Selected Remote Control Schemes Installed in Home

3.4.5 Further Discussions

Only remote control aspect of home automation system is considered here for its betterment and easy implementation. Essential schemes for remote control of appliances are proposed here and depending upon the circumstances and requirements one or more of the schemes

can be chosen for installation in a given system. Relay register and relay board unit is the essential hardware unit present in the home automation system where selected remote control schemes could control the bits in relay register. Whatever may the control technique used in the system the final control element would be the relay operating to switch ON or OFF of the selected appliance. Internet based control performed through web is versatile and interactive in controlling the appliances. If more and more appliances are needed to be controlled then an encoder is needed to be used in the transmitter to have the standard word length of 16-bits for the control word. In the receiver the 16-bit control word is decoded and the corresponding bits are taken to the relays for switching the appliances.
In simple automation system where internet facilities are not provided one can opt cellular phone based control scheme which is simple and cost effective. Alternatively for such requirements landline telephone with an extension card could also be opted.

3.5 Personal Digital Assistant

The personal Digital software is installed in PC which is linked to the operating system that brings the daily chart to the screen of the PC monitor. In the database accessible to the OS, the various activities to be done on different dates and times are pre-programmed and this updated daily by the user. The personal digital software refers the home data base every morning at the set time and brings the list of activities to be done on that day to screen as to alert the user to be ready for solving the issues of the day.

4. Home Security Concerns

An important home security aspect insists that the fire accident that can happen in home during the absence of the resident should be alerted and intimated to a remote location. Also intruder making unauthorized entry into the flat should also be informed to a remote location for taking follow up actions. Authorized visitor entry supported by video image presentation is another aspect which helps in security concerns.

4.1 Fire and Intruder Alert

4.1.1 Fire Detection and Intimation

The occurrence fire in home is detected electronically. It is then sent as SMS and then converted into email and sent again to a remote PC. The arrival of email in remote PC executes a file to create an alarming sound. The email can also be received in a cell phone causing a beep sound. Also the detection of intruder by a sensor sends another SMS which is converted into email as to make appropriate indication in remote PC or in cell phone.

4.1.1.1 System Overview

A simple scheme sending SMS upon fire detection or intruder detection and then its email conversion is shown in Fig.8. An 8085 based microprocessor system is extended to light and heat sensors through interrupt interface. Through I/O ports of the microcomputer a GSM module is connected. The GSM module is backed up with the facilities of TynTec system (Tyntec website, 2009). If excessive light or heat is sensed electronically in the flat and if any of its voltage exceeds the threshold then it interrupts the microcomputer (RST7.5). The

interrupt service subroutine is arranged to send a text message to GSM module which in turn sends an email to the number provided by TynTec. The two way SMS online tool of the TynTec converts the SMS into an email and forwards to the specified email-id. This email message can be received in an internet backed PC. As described before the email platform is set up with a rule such that as soon as the email arrives with a specific word in the text the *exe* file would run a set of instructions as to give alarming message. The email can also be received in a cell phone causing beep sound alerting the user. The detection of intruder interrupts the microprocessor on the interrupt line RST6.5. The interrupt service subroutine of RST6.5 sends the related message to GSM module and this in turn sends the email to the concerned email-id.
The detection of fire is made by sensing the exhaustive heat or light electronically and comparing with a threshold. The sensor circuit is shown in Fig.9. In place of Rx, LDR is used for light sensing and thermistor is used for heat sensing. These sensitive elements are kept at selected spots in the home. Rx might be a single element or an array of elements where all elements representing Rx are connected in parallel to the sensor circuit. If anyone becomes active it pulls down the effective resistance of Rx resulting in increased voltage to the comparator.

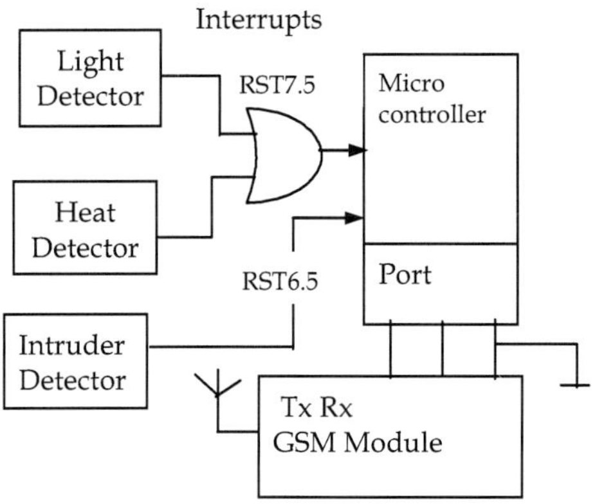

Fig. 8. Simple Schematic of Alarming by Fire and Intruder Detection

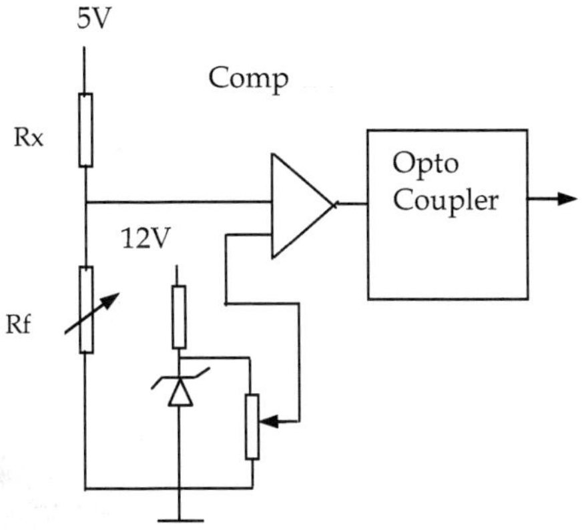

Fig. 9. Sensor Circuit

4.1.1.2 Fire Detector Actuates Telephone dialling

As an alternative to SMS and email alert, the fire detector can be arranged to actuate dialling through landline phone kept in home to the resident's mobile phone. An electronic dialling circuit attached to home telephone can be triggered by exerting actuating signal obtained from any type of detector circuit (Balasubramanian and Cellatoglu, 2008). The actuating signal given to it does two things; *i.* it operates a relay circuit to close the cradle for a pre-determined time so as to receive dial tone and to commence dialling, *ii.* it provides enable signal to a sequential logic circuit that injects the series of pulses corresponding to chosen digits into the telephone lines. We now use this facility to dial the digits of resident's mobile number whenever fire detector is activated. Upon detecting fire it actuates the circuit to give a call and the call reaches the resident's mobile giving ring. Instantly, the resident knows that call arrives from home and understands that there is fire in his home. He immediately dials back to home telephone with additional digit as described in section 2 where the telephone has a relay board extension to control selected appliances. Receiving the additional digit in home telephone is arranged to switch ON a relay circuit as to operate a solenoid valve triggering a fire extinguisher electromechanically. The fire extinguisher is kept at chosen location inside the flat. The fire extinguisher can be set to operate manually also. If in case the resident is inside home and receives the call from his home telephone in his mobile he can opt to extinguish fire manually or by giving ring.

4.1.2 Intruder Detector and Intimation by SMS

In a similar way of implementing the unit of heat detection and intimating the intruder detection and intimation is also made. In the sensor circuit (Fig.8) the places of LDR (Rx) and Rf are interchanged such that normally the LDR receives full light from a light beam. When an intruder interrupts the light falling on LDR the voltage given to comparator raises and exceeds the threshold. The interrupt service procedure of the microcomputer is prepared for sending a message for intruder detection. Consequently the SMS and email are sent accordingly as explained before.

4.1.3 Updating Control Table Flags

As mentioned already the two most significant bits of the status register will be updated by the fire detector and intruder detector as and when they get activated and the flags can be seen in the control table (Table 1). The flags will remain for 6 hours period and if desired it can be changed to any other desired value with the monostable multivibrator.

4.1.4 Discussion

The intruder alert and fire alert given automatically to a remote location through SMS from mobile phone converted into email is a passive system passing the message to the concerned for taking counteractive measures. The message can be reached through landline phone as well. One counteractive measure would be to dial the additional digit to operate the fire extinguisher. On the other hand, the fire extinguisher could be included in the list of appliances for web control if desired. In this case the user can also see the status register to understand the present status of fire detector or intruder detector and take action through web control table itself. All these techniques presented here are expected to enhance the remote control and alerting features of home automation systems.

4.2 Regulating Visitor Entry

The regulated visitor entry unit helps checking the visitor through video and admits the intended visitor to flat. A preliminary form of video door entry was reported in the past (Balasubramanian et al, 1999). We enhanced the performance of the system (Balasubramanian and Cellatoglu, 2008) in certain aspects and now more considerations are given. This regulated visitor entry unit has a register to record the visitor's time and date. Maintaining time and displaying in monitor screen would be useful to the resident and also for recording in the visitor register.

4.2.1 Visitor Admittance and Recording

Home Security is an important aspect of home automation. Permitting the authorized visitors to flats is a main concern to home security. The visitors seeking entrance into a desired flat has to be checked and admitted as to ensure the security measures. A special I/O card having several I/O ports, Interrupt interface circuits and Timer are interfaced to the PC inflate. Fig. 10 shows the simplified schematic of the I/O card.

When the visitor comes near the main entrance gate, he depresses the button of the concerned flat which opens the main gate, if not open earlier by others, and triggers the video system to switch ON all cameras. This action also gives a visual indication in the display panel of the flat. Momentarily the mini video monitors in all flats get ON and show these images. The images of the video cameras are displayed in four quadrant windows in the monitor screen. One of the quadrant in screen shows the image of the video camera kept at the door of the flat. The video images of all staircase cameras are sequentially surfed on the screen. Also, the status of the video system ON/OFF is informed to the PC through an input port for making it prepared for entering in the visitor register. The ON period of the video system is programmed as five minutes and within this period if another visitor arrives then it would continue for five more minutes. The resident watches the video monitor, particularly the image of the camera kept at the door. When the visitor arrives the door, the door is let open by the resident upon his decision.

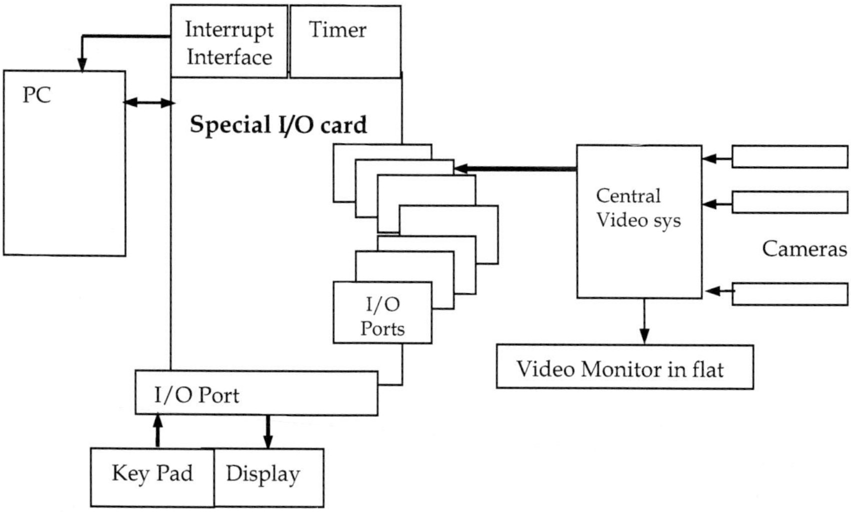

Fig. 10. Simple Scheme of Visitor Admittance and Recording

4.2.2 Visitor Registor

A file is organized in hard disc to have a record of the visitor with time and date taken from the *time and date register*. When the door is open by the resident this also gives an interrupt to open the visitor register and creates a new record. The new record in the file is automatically loaded with the serial number and the date and time fields. It is up to the resident to enter the name or identity of the visitor as a window slot that would appear on the screen for making entry. The opening of the door lock by anyone using the key does not activate the interrupt and hence no changes are made in the visitor file. This avoids recording the resident himself as visitor each time when he enters to the flat. The file can be viewed by the resident at any desired instant of time when needed.

4.2.3 Time and Date Register

As in the past (Balasubramanian, 1991) a digital clock is maintained by the PC by organizing software counters for seconds, minutes and hours. RAM memory locations in PC are organized as software counters. The timer of the I/O card (Fig.10) produces pulse train at the rate of 1 Hz and this is used to interrupt the system. In the interrupt service subroutine the seconds counter (MOD-60) is incremented and the result is carried to minutes counter (MOD-60) and also to the hours counter (MOD-24). These registers are initialized and updated from the system clock available with the Operating System. There is reset option given by the user as to initialize zero time necessitated to run a stop watch for observing time count of specific events. In addition to displaying the time and date information in PC monitor they are tapped out to make display in large sized LED panel, if required.

4.3 Security Threats and Countermeasures

As internet and email are used as remote control methods of controlling the appliances there is every chance for security threats to occur. The threats occurring through internet such as virus can be overcome by installing appropriate antivirus software. The use of commercially available home automation control devices may interfere each other due to mismatch in protocol and might cause interruption in their services. Since the proposed architecture uses commonly available relays its functioning being clear it doesn't encounter such difficulties.
The monitoring program can be periodically reactivated and updated by the concerned software package available with the user. This can accommodate including additional devices included in the smart home.

5. Mini Electricity Generation Unit for Home

Cost effective and compact electricity generation units depending on renewable energy resources has been reported in the past (Balasubramanian et al, 2009). A mini electricity generation unit working with automatic direction controlled solar panel and mini wind mill. This extracts maximized energy from the solar and wind energy. The voltage developed in the solar panel is used to charge an accumulator as to preserve the electrical energy. The AC voltage from windmill is rectified, boosted and also saved in accumulator. The accumulator voltage is inverted to have regulated frequency and delivered to home. This unit satisfies some percentage of home electricity consumption.

5.1 Control Unit and Generation of Electricity

The dual energy system extracting the electrical energy from the wind and solar resources is shown in Fig.11. The windmill has a tail part in it which aligns itself to the direction of the wind as to extract the maximum energy from the wind and no electronic controller is needed. The speed of rotation of the blades and hence the voltage output from the generator will be maximum if the plane of the blades is facing normal to the direction of the wind. The AC voltage generated is boosted to a higher level with the help of booster converter employing inductor diode and electronic switch (FET) controlled form the microcontroller. Based on sensing the voltage of the windmill the microcontroller arranges PWM pulses for controlling the duty cycle as to boost the voltage level. Through a well designed charging circuit it charges the chain of accumulators to save the energy extracted from wind in terms of DC voltage.
The DC voltage from accumulator is converted into AC voltage of the standard of the country and is extended to the residential electrical load through dual direction (net metering website, 2009) running energy meter. If windmill is working at its full speed generating 1KW power and if the residential appliances are not taking this power then the power will be fed to the power grid making the meter to rotate in other direction selling energy to the grid.
The rotating speed of windmill depends on wind flow rate. The frequency of the AC voltage and power depends on wind flow. As the wind flow is non uniform, the frequency of the AC is fluctuating with the wind speed. This AC cannot be directly taken to the household appliances with varying frequency. Therefore, AC is converted into DC and saved in accumulators. After then it is inverted to AC again by switching DC into 50Hz rate as to make the standard 50Hz frequency useful for household appliances.

There are other ways of stabilizing the speed of the windmill under varying wind flow conditions (windmill website, 2009). But this needs additional control hardware resulting in increased cost.

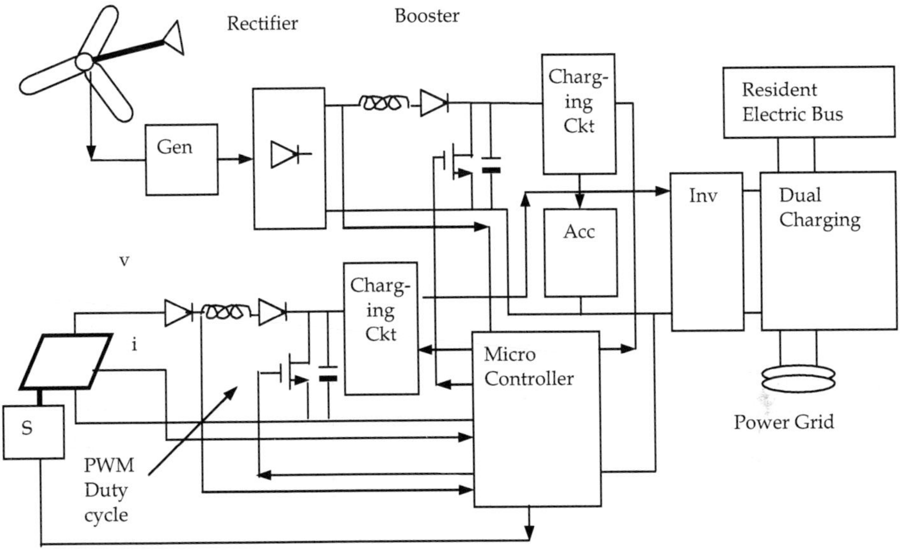

Fig. 11. Dual Enegy System Controller

5.1.1 Solar Tracked PV Panel and Optimal Power Tracking

The primary tasks of microcontroller are *a.* chopping the booster converter *b.* optimal power tracking with the PV panel and *c.* solar tracking. Parallel to charging the accumulator from the wind power the boosted PV array power also is charging the accumulator. The PV panel is kept always aligned to the direction of the sun as to maximize the voltage output. In order to do so, it is fixed with LDRs through hollow cylindrical tubes kept at all corners and also in the middle of the panel. The resistances of the LDR depend on the exposure of light at the angle of the panel with respect to the direction of the sun. The voltages are picked up in bridge circuits and taken to the microcontroller where a decision is taken to rotate in orthogonal directions by two stepper motors. The voltages picked up in the PV cells of the solar exposed PV panel are added up and the cumulative voltage is read by the microcontroller. In order to boost the PV panel voltage an inductor is connected to it and is switched ON periodically by the microcontroller. The boosted voltage is fed to the charging circuit of the accumulator. There is another charging circuit associated with the windmill and the voltage levels of the two charging circuits are kept as the same. During day time both chargers would work in parallel and during night time the wind mill charger might alone work.

The earlier sun tracker algorithms (Koyuncu & Balasubramanian, 1991; Balasubramanian & Cellatoglu, 2009) are updated for present application. The voltages due to left, right, top, down and central LDRs are obtained external to the microcontroller and they are read and processed. The differences in the left and right are computed and drive signal is issued to

one motor. Likewise, the difference in voltages due to top and bottom LDRs is computed and drive pattern is obtained for driving the other stepper motor.
The sun tracker algorithm and the hardware arrangement keep the panel rotated from east direction to west direction all along the day tracking the sun. After sunset, with a delay of a predetermined period of time, the panel is brought to east direction step by step as to keep it set ready for facing the sun next day. The sun tracker algorithm and power tracking algorithm are performed simultaneously in an interleaved manner.

6. Conclusion

There are several smart home devices with their control units available in practice. The appliances are connected to the concerned bus and the switching is monitored with their software. For instance X10 devices (website 2009) which are used in communication protocols are employed in home automation applications. This needs the x10 transceivers to be installed at desired nodes. This project does not rely on any commercial smart home devices and the control is effected by simple means through relays which are commonly known to everybody. In case of arising any fault performing the fault diagnosis is quite easy.
When smart home devices of different makes are used there is a quite possibility of mismatch to occur when control actions are effected from common software. When internet is involved in effecting control actions to devices of various makes the security threats are more vulnerable. In the proposed relay based control the security threats could be easily handled.
Based on the information provided here, one can build the system incorporating the facilities required for his flat in the building. We don't normally apply the controlled switching for all electrical appliances in the flat. When the resident is away he would like only a very few devices to be operated as to serve certain needs arising at any moment. Therefore, we considered only selected devices to be operated under controlled action besides the manual switching facilities available for the same.
Keeping the requirements of modern home automation system accommodating most desirable features the home automation system has been designed and reported. The selected appliances to control by remotely through phone and internet and also by local means are taken as example only. Although, it satisfies most requirements of consumers, if in case the user wants to control few additional devices it can easily accommodated.
Timers associated with the monitor program installed in the PC can be programmed to switch on cyclically the selected appliances for security concerns. For instance, the periodical illumination of lamps by a timer would simulate the condition of resident staying in the flat.
The solar panels and windmills concede a part of electrical energy requirements of the flat. Since the panel and windmill are kept on the roof their clear exposure to sun and wind would extract maximum energy and this is aided by direction alignment act. The extraction of energy from these resources becomes important nowadays due to the shortage of main resources such as fossil fuels.
The security measures are efficient with the authorized visitor entry, control of appliances and intruder detection. The presentation is made in a simple way that one can easily follow and develop and the technology is open for anyone to accommodate more features by adjusting the hardware and software.

7. References

http://www.homeautomationforum.com, 2009.

http://www.home-automation.org/, 2009.

Jorge Caleira Nunes and Sérgio Jardim da Silva, "Adding intelligence to home automation systems", http://www.iadis.net/dl/final_uploads/200401C044.pdf

Renato Nunes, "Demobus-A New Approach to Home Automation", *Proceedings of 8th International Congress on Electrical Engineering*, Portugal, July2003.

K. Balasubramanian and A.Cellatoglu, "Improvements in Home Automation Strategies for Designing Apparatus for Efficient Smart Home", *IEEE Trans. on Cons. Electronics,* Vol 54 , No 4, Nov 2008, pp 1681-1687.

http://www.mobile-phone-directory.org/Glossary/C/Cellular_Phone.html 2009

K. Balasubramanian, "Control appliances remotely via the telephone", *Electronics World,* Vol 109, No 1809, Sep 2003, pp 20-21.

http://www.tyntec.biz/ 2009

K. Balasubramanian and A.Cellatoglu, "Intruder Detection and its Remote Intimation by Autodialling", *Proceedings of the First International Conference on Emerging Technologies and Applications in Engineering, Technology and Sciences,* Rajkot, India, Jan 2008, Vol–I, pp 550-555.

K. Balasubramanian, H.Camur and Rajaravivarma, "Microprocessor based Video Door Entry System: A Supplement to Home Automation", *Abstracts of the 1999 IEEE SoutheastCON Conference,* Lexinton, Kentucky, USA, Mar 1999.

K. Balasubramanian, "Multi-time zoned digital clock", *IEEE Trans on Cons. Electron,* Vol 37, No 4, Nov 1991, pp 867-872.

K. Balasubramanian, A.Yasli and A.Cellatoglu "Dual Energy System Design for Wind and Solar Resources Meeting Residential Energy Needs in North Cyprus", *Proceedings of the National Conference on* Emerging *Trends in Engineering Technology and Applications (NCETETA-2009),* Bangalore, Apr 2009, pp 391-396.

http://apps1.eere.energy.gov/states/alternatives/net_metering.cfm , 2009

http://en.wikipedia.org/wiki/Wind_power, 2009

B. Koyuncu and K. Balasubramanian, "Microprocessor controlled automatic sun tracker", *IEEE Trans on Cons. Electron,* Vol 37, No 4, Nov 1991, pp 913-917.

K. Balasubramanian and A.Cellatoglu, "Optimized solar panel and wind mill alignment system for full utilization of solar and wind energy for consumer use", Proceedings of the International Conference on Energy Engineering, Pondicherry, India, (ICEE-2009), Jan 2009.

www.x10control.com , 2009

4

Applying agent-based theory to adaptive architectural environment – Example of smart skins

Shang-Yuan CHEN[1] and Shu-Fen CHANG[2]
[1]*Feng Chia University*
[2] *Ming Chi University of Technology*

1. Motivation and Goals

Establishing a sustainable environment is an important mission for research on the next generation of intelligent houses. The current living environment is full of fixed architectural elements, or reactive elements that merely possess a "perception — reaction" function. These elements are hard to respond to a changing and complex environment by judging and assessing changing circumstances in order to satisfy users' varied, complex needs. Take an opening in a classroom wall, for instance: When loud construction noise from outside can come in through the opening, but the classroom is extremely hot because of crowded students, would it be better to open or close the window in order to achieve a comfortable classroom environment? Of course this situation cannot be analyzed solely on the basis of use of a fixed glass soundproof door or window, or a ventilation opening, and can be resolved by a reactive device possessing solely a perception — actuation function only to a limited extend. As a consequence, this study seeks to promote the creation of an intelligent architecture environment possessing perception—computing—actuation—communication (PCAC) functions, where the computing mechanism can consider diverse situations, drive architectural elements, and engage in communication and coordination in order to adapt to changes.

From the point of view of "context awareness," adaptivity implies the possession of smart entities able to sense and respond to changes in external circumstances, adjust internal functions, and independently adopt actions meeting current needs. In addition, a building's possession of adaptivity implies that, apart from making reasonable assumption, the smart entities will have enhanced ability to communicate and interact with users. Context implies situational information; system possessing context awareness can extract, interpret, and use contextual information, adjust its functions, and provide applications for users or to accomplish its mission (Selker, 2000; Dey, 2000). From the bottom up, all buildings contain numerous switches, and these switches - like brain cells - can be used to construct a logical system. From the top down, buildings and their environs contain contextual information concerning who, what, when, and where. How should we apply this information, and for what purposes should we use this information? In order to obtain and apply effective contextual information, a building must include an intelligent entity capable of processing information and inferring its meaning. Table 1 lists the dimensions and content of the "six W's" for contextual information concerning building environments.

Why	Reason for system design and goals
HoW	How to apply contextual information - steps and methods
Who	1. User identity — sex, age, characteristics, occupation 2. Physiological measurements — blood pressure, heart rate, breathing rate, muscular activity, tone of voice 3. Psychological preferences — like, aversion, happiness, anger 4. Types of activity — conversation, reading, walking and running 5. Social situations — company, group, status, jurisdiction
What	1. Variable factors affecting adjustment of system testing and assessment needs 2. System-related equipment, devices, servers, and spaces 3. Effective resources — batteries, displays, networks, and bandwidth, etc.
When	1. Temporarily-stored information — preserved for a certain period of time, such as one day, one quarter, or one year 2. Appointments — itinerary, agenda 3. Event-driven opportunities, event continuation time, and event recurrence period
Where	1. Spatial information — location, direction, bearing, speed, acceleration, motion 2. Environmental characteristics — temperature, humidity, illumination, air quality, wind speed, wind direction, noise, rainfall

Table 1. Contextual information six W dimensions and content

This study seeks to employ intelligent agent theory to investigate an adaptive architectural environment, takes smart skins as a research example, and proposes the use of neuro-fuzzy to establish judgment and reaction control conditions. Via a literature retrospective, issue research, case analysis, and summarization of conditions, this study derives adaptive environmental research categories, establishes adaptive building environment hypotheses based on intelligent agent theory, summarizes and analyzes the numerous feasible computational mechanisms and selection conditions, and employs a prototype smart skin structure to design experiments, perform testing and assessment, and analyze the relationship between users, the environment, and the smart skin.

2. Research Scope and Content

The smart skin hypothesis proposes that is a building envelope possesses intelligence, it should be able to simultaneously take the needs of both users and environment into consideration. Here environmental considerations include the life-related functions of the outdoor environment (sunlight, wind, noise, and rain, etc.) and indoor environment (temperature, humidity, and artificial lighting, etc.). User-related considerations include for the need for both physiological and psychological comfort (Fig. 1).

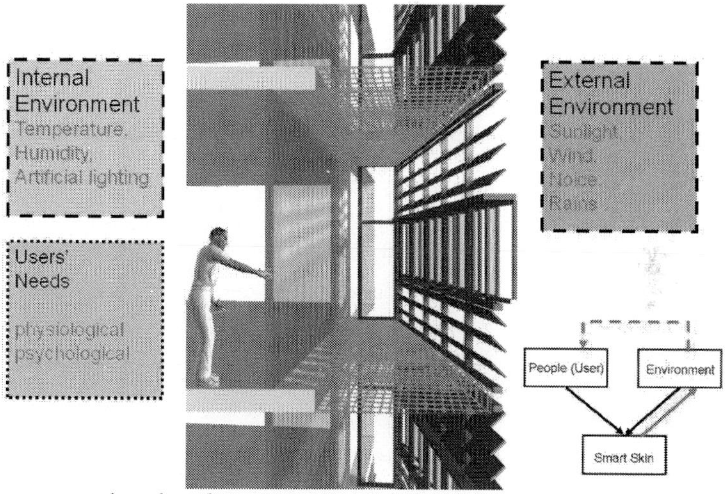

Fig. 1. Things to consider when designing a smart skin

This study employs intelligent agents as a framework for intelligently integrating people, objects, and spaces. These agents possess the function of context awareness, and fuzzy theory enables the agents to find optimal solutions among numerous possible computing mechanisms. From a macro perspective, the adaptive mechanism research categories derived in this study include smart house, agent theory, and context awareness (Fig. 2).

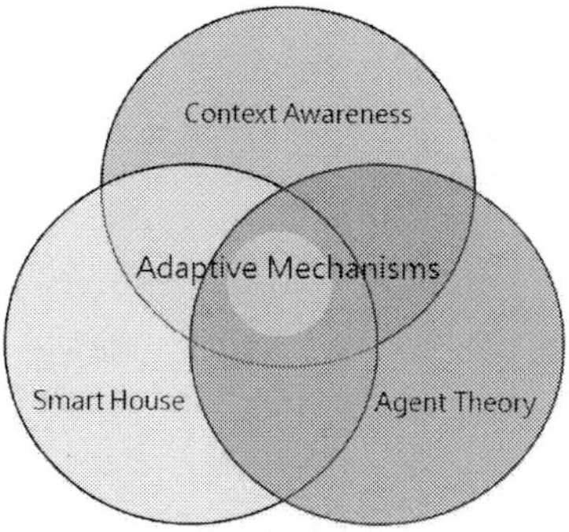

Fig. 2. Adaptive mechanism research categories

From a micro perspective, the core content of adaptive mechanisms is integrated on a platform of intelligent agent theory (this content includes context awareness, and interface and database design). Dimensions that must be taken into consideration include input perception conditions (including information concerning users and the environment), output action models (taking a smart skin as an example, these include the skin's characteristics and changes in the composition of each level), and feasible computing mechanisms to perform processing and assessment (including rules and neural network and fuzzy theory; the computing mechanisms derive and select optimized adaptive effects and actions on the basis of input and output conditions). (Fig. 3)

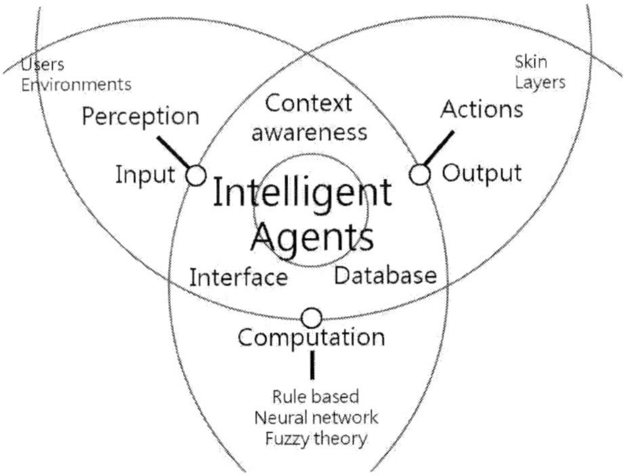

Fig. 3. Content of core research on adaptive mechanisms

3. Literature Retrospective

The smart house concept is derived from a series of transformations in dwelling technology. Due to the electrification of homes in the early 20th century, the availability of clean and convenient energy, and the use of household appliances and other applications, initiated a transformation in dwelling technology. At the end of the 20th century, the introduction of information and communications technology, and especially the Internet, into the household created a host of implications that are still being explored today. Intelligent agents are an important current research direction in the field of artificial intelligence. In an environment with distributed intelligence, computing, information, and communications mechanisms serve as tools for representing knowledge. Recent research on smart houses has incorporated sensing technology, computing technology, and information and communications technology in order to bring about self-programming ability better able to reflect users' living habits (Mozer, 1998), and has attempted to achieve zone and dispersed control mechanisms on the basis of past central control models (Junestrand, 1999).

In an agent-based living environment, researchers, designers, have long been perplexed about how to select appropriate technologies, and are uncertain how to deal with these technologies. For example, the rule-based computing system employing binary logic used in the Sentient Building at TU Vienna (Mahdavi, 2005) employs a dispersed, hierarchical control node structure, where the nodes constitute information processing and decision-making control points. As a consequence, more meta-controllers must be added as the number of devices increases. The fact that it is not easy to distinguish modules elements in the system increases the difficulty of control and rule description. In another example, the Adaptive House (Mozer, 2005) employs a central neural control system termed the "Adaptive Control of Home Environment" (ACHE), to strike an optimal balance between maximum user comfort and minimum energy consumption. However, assessment of this system has found two main problems causing the neural system converge on a state of low energy consumption and low comfort: The first problem is that the system's X-10 controller is often slow to respond or not working properly, and the second problem is improper user operation. As a result, the system tends to deteriorate, causing the central neural computing system to perform erroneous learning.

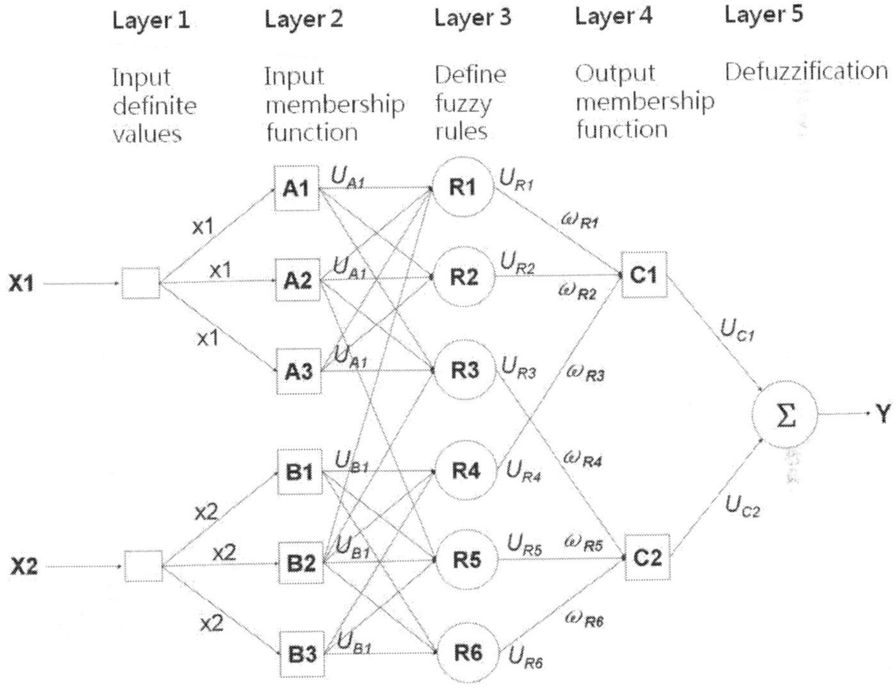

Fig. 4. A Neuro-Fuzzy System

After examining the foregoing cases, this study decided that the selection of an agent computing mechanism select must simultaneously take into consideration the three aspects of the situation, computing mechanism theory, and hardware and software technology. The study therefore proposes the use of a neuro-fuzzy concept combining fuzzy logic with a neural network as the agent computing mechanism. This approach pairs human logic with rational learning and adaptation ability. A neuro-fuzzy system employs fuzzy rules in the form of associated weights, which projects the neural network structure on a fuzzy logic system, causing the fuzzy logic system to possess the learning algorithm functions of a numeral network. Because of this, a neuro-fuzzy system is able to allow a smart skin to change or adjust its rules on the basis of sampled user experience-based information. In other words, a neuro-fuzzy system uses the steps of (1) fuzzification: input of clear values and a membership function, (2) definition of a fuzzy rule base, (3) fuzzy inference: output of the membership function, and (4) defuzzification to create a quasi-multilayer back-propagation neural network structure. Neuro-fuzzy learning relies on training by example to adjust the associated weights constituting the fuzzy rules. Fuzzy associative memories (FAMs) are fuzzy rules possessing associated weights. Altrock (1995) defines associated weights as degree of support (DoS), where degree of support expresses support for that fuzzy rule. The maximum value of degree of support is one. A neuro-fuzzy network employs an error back propagation algorithm, and adjusts degree of support to correct the error between the result obtained using the original fuzzy inference rules and the actual output value, and thereby achieve an optimal correspondence. Fig. 4 shows the neuro-fuzzy system, where ωR1-ωR6 are degree of support (DoS) values. (Negnevitsky, 2005)

4. Establishment of an Agent-based Smart Skin

A smart skin is defined as a building envelope that is able to perform adaptive intelligent activities by changing its skin and layers (including via reaction, action, interaction, and communication) following computing and inference based on perceived effective external information, and can thereby satisfy users' needs for comfort and environmental sustainability. As far as perception factors are concerned, effective information is derived both from the environment - "the place" - and from the internal users. In addition, a smart skin also depends on hardware and software systems comprising sensors, computing equipment, and the building's actuating elements to achieve perception - computing - actuation - communication context awareness functions. The main environmental factors and variables operated on by the driver agent-based intelligent objects are analyzed below:

Information from the environment and place can be classified as indoor and outdoor information. Outdoor information includes such items as light, noise, heat, air, moisture, and view. Indoor information includes illumination, temperature, humidity, security, and health. Effective user information includes psychological and physiological items; psychological information includes happiness, likes/dislikes, privacy, preferences, and respect; physiological information includes location, posture, age, sex, glucose, heart rate, and alone/with company. Adaptive actions of the outer shell can take the form of changes in the skin or in different layers.

Adaptive actions performed by the skin can be classified as performance changes and movement changes. Performance changes include changes in appearance, material, color, thickness, density, pattern, or mixing and matching. Movement changes include changes in opening method (such as changes in opening shape or size), translational motion, movement, rotation (angular change), and change of degree (such as change in transparency or density).

Adaptive actions performed by different layers can be classified as composition changes and layer changes. Composition changes include addition (variety and diversity), reduction (minimalist style), multiplication (repetition and differentiation), and divided (modules and elements). Layer changes include single-layer and multiple-layer changes, the relationship between the support and infill, the arrangement of skin layers (upper, middle, and lower or inner, middle, and outer), and the relationship between skin layers and the building mass (such as adhesion, incorporation, and separation).

In addition, with regard to hardware and software facilities, apart from consulting the content of a contextual knowledge base containing the foregoing perception and action information, the installation of sensors and actuators must also take into consideration the distribution, delineation, and density of sensor and actuator hardware, and their times of action, such as continuous actions times, intermittent action times, and action period settings (Fig. 5).

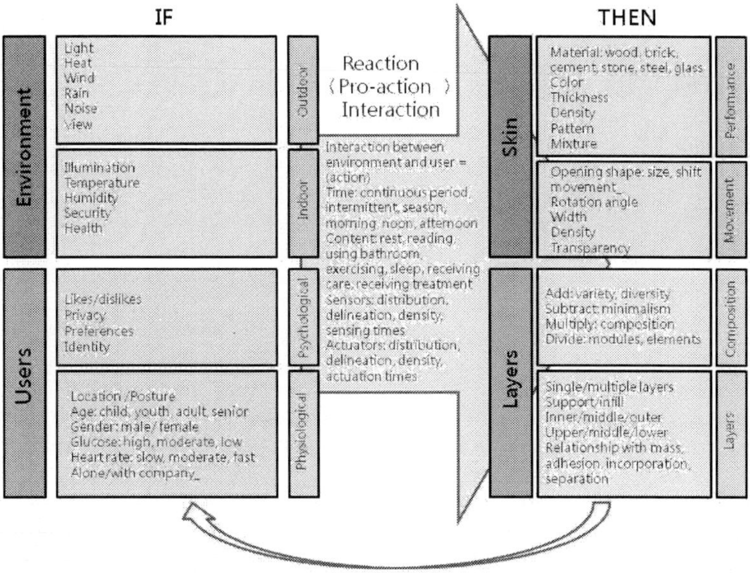

Fig. 5. Model of a smart skin framework with user-oriented context awareness functions

In short, the basic elements of a smart skin consist of sensors collecting external information, processors performing computing and inference, and actuators (architectural elements) outputting movements. A smart skin can change and adjust the state of the skin in accordance with changes in the external environment in order to maintain optimal user comfort and environmental sustainability.

4.1 Use of intelligent agent theory as an integration framework

An agent-based control system can be divided into two parts responsible for describing and setting the responses and actions of intelligent devices. The first of these consists of an independent intelligent agent module and its computing mechanism and plans, and the second consists of the intelligent agent community and its interaction model.

Software agents are able to perceive the environment and choose an action to implement to influence the environment (Russell, 2003). So-called perceiving is performed by sensors that receive information from the environment, and so-called actions refer to the agents' ability to influence the environment. Agents must be able to react promptly, and must also work proactively to achieve their goals. The key to balancing action and reaction lies in the changing situation; specific situations can be referred to as events (Fig. 6).

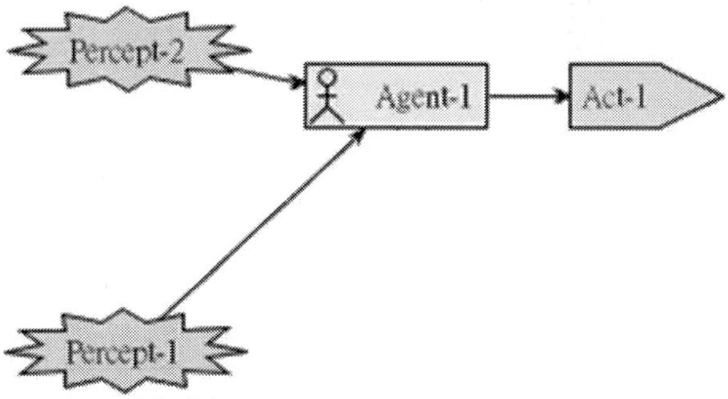

Fig. 6. An Intelligent Agent Module

Plans and sub-plans must be drafted to ensure that the system can effectively achieve its goals; these plans and sub-plans describe the cause and effect relationship between perceived events and output actions (Padgham, 2004). As a consequence, each agent's basic module is composed of sensors, computing mechanisms, and actuators, including software and hardware (Russell, 2003). Software agents process information received from sensors or other agents via an event-driven model, and then drive the building's in-filled components in accordance with plans or sub-plans, and perform reactive, proactive, and interactive adaptive behaviors. (Padgham, 2004). Reaction refers to immediate action taken by an agent without computing after receiving information. Proaction refers to action taken following computing after receiving information. Interaction refers to communication between an agent and other agents or a person via an interface (Fig. 7).

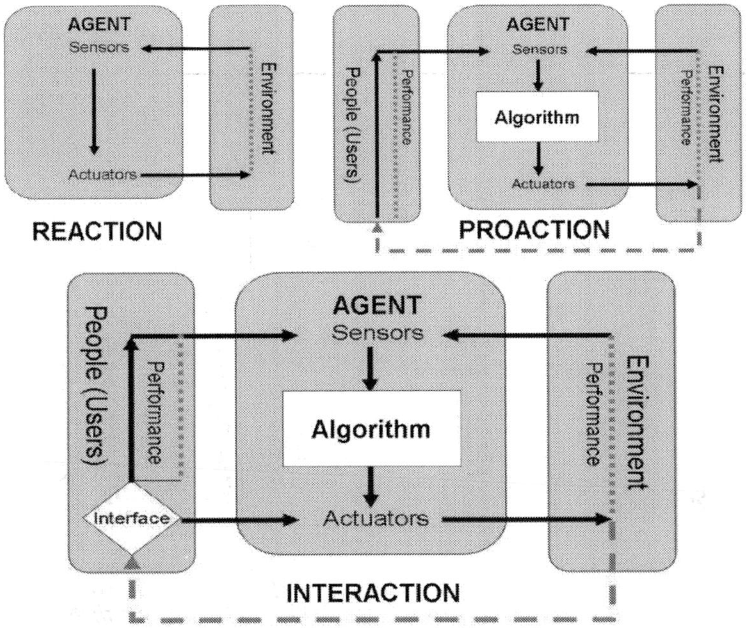

Fig. 7. Adaptive behaviour by intelligent agents

Agent communities can generate cooperative or coordinated interactive behaviors (including one-to-one, one-to-many, and many-to-many relationships) via common communications protocols, shared databases, messages, and messages transmitted by agent communities (Wooldridge, 2002). The levels and subordination relationships of agents within a community may change as they are reassembled to suit a goal or mission (Minsky, 1988) (Fig. 8).

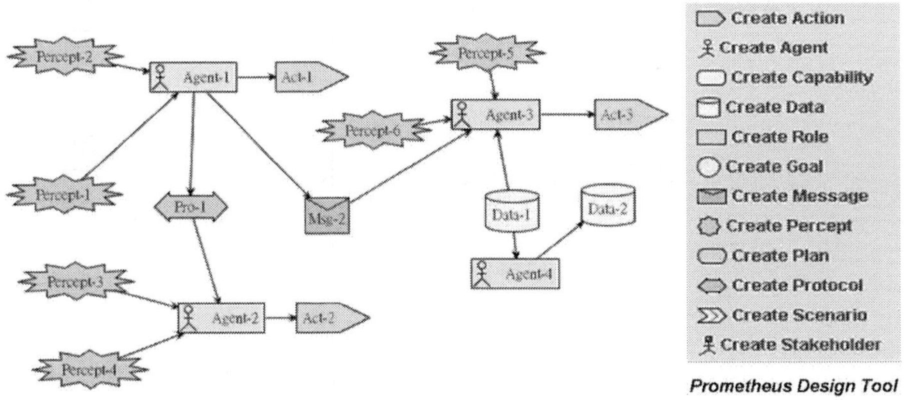

Fig. 8. Interactions in a community of intelligent agents

4.2 Existing technological conditions

An agent-based smart skin requires three main elements: sensors, a computing device, and actuators. A data logger (CR510, Campbell Scientific Canada Corp., 2007) is a feasible computing device; this data logger is a data acquisition center, and is able to receive data from most sensors and allow program design (Fig. 9). Using the data logger as the computing core of the smart skin, data processing proceeded as follows:

Input signal from sensor <−> data logger <−> network server <−> output to actuators

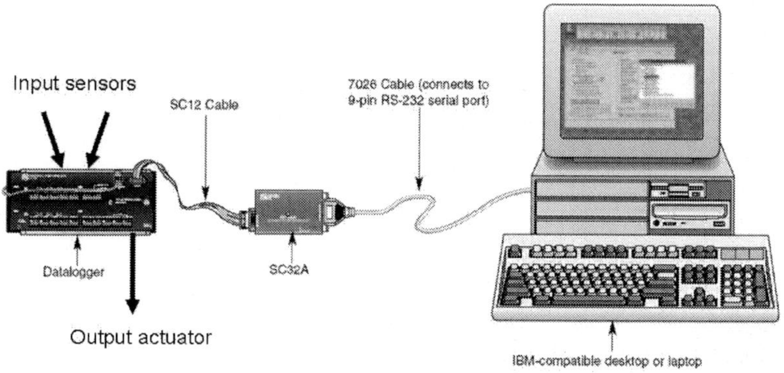

Fig. 9. CR510 data logger (Campbell Scientific)

The start of measurements and control of functions are based on time or event. The data logger is able to drive external devices, such as pumps, motors, alarms, freezers, and control valves. The data logger's program software is known as EDLOG. EDLOG contains four processing elements: (1) input, (2) processing, (3) program control, and (4) output processing. We can therefore infer that the smart skin's processing flowchart will be as shown in Fig. 10.

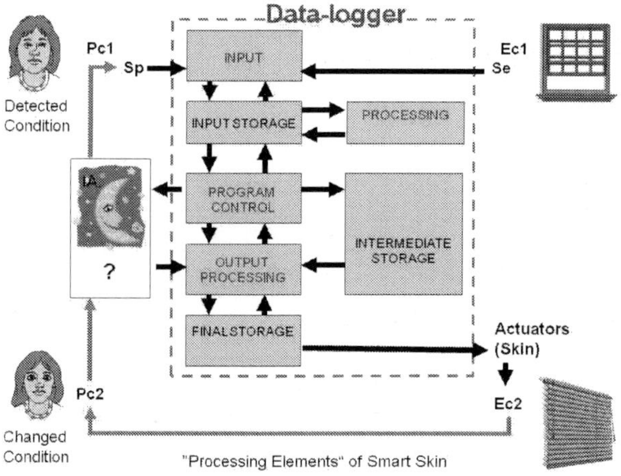

Fig. 10. EDLOG's four processing elements and smart skin processing procedures

Fig. 11 shows an example of the EDLOG program's plans. In addition, apart from the core program, because the system also required an agent interface design, executable files in the VB programming language were to activate interface agents. Database applications programs (Dreamweaver+ ASP+ Access) were used to design a user interface and establish a database. The establishment of a database involved the storage of user class data, and environmental change history and smart skin interaction records.

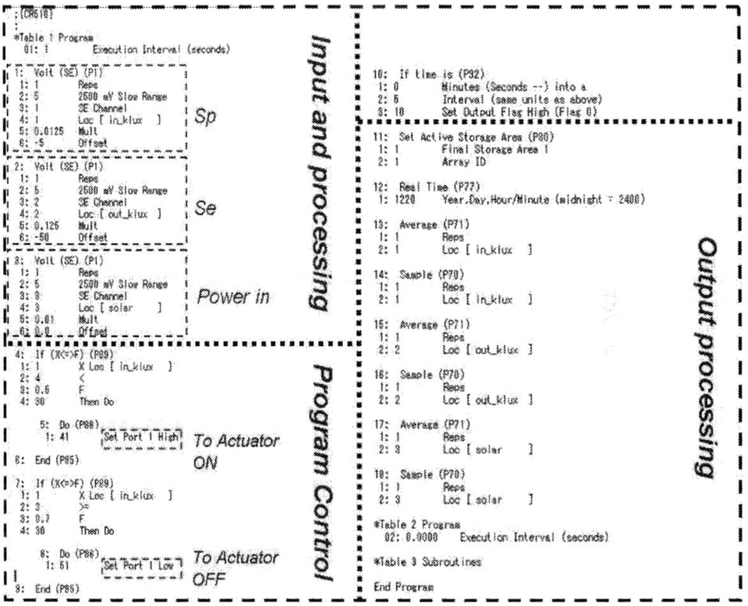

Fig. 11. Example EDLOG program

The smart skin modelled using the data logger verified the feasibility of developing an adaptive architectural environment on the basis of intelligent agent theory. In accordance with the foregoing analysis, the use of a binary logic rule-based computing mechanism possesses the following advantages, which make it easy for people to understand and allow it to reuse knowledge: (1) It can readily represent natural language knowledge; (2) it possesses an IF-THEN format structure; (3) it can easily extract knowledge from the problem solving process; and (4) it can employ "EQU", "AND", and "OR" statements to express agent-based adaptive behaviour. Nevertheless, rule-based computing mechanisms have the following major disadvantages, which prevent from being the main computing mechanism for agents: (1) The restrictions of rule-based logical conditions limit learning from experience. (2) While "AND" and "OR" binary logic can resolve conflicts where compromise is possible, they cannot resolve conflicting "XOR" situations; this necessitates the use of higher-level decision-making and control mechanisms, and prevent these mechanisms from being independent smart modules. (3) The binary logic lacks the ability to express multiple values and continuous values, which makes it difficult to resolve complex problems.

This study recommends that a neuro-fuzzy system be used as the computing mechanism for an intelligent agent module, and user-friendly fuzzy inference and neuro-fuzzy learning technology be used to establish an adaptive user experience-oriented building environment. In comparison with other adaptive technologies, neuro-fuzzy has the following advantages: (1) Because the system is constructed on the basis of fuzzy logic, learning freedom is controlled, and erroneous learning is avoided. (2) The system inherits knowledge from fuzzy logic systems, and can therefore interpret or make inferences from the results of learning. While smart skins with rule-based reasoning ability lack the adaptive ability needed to respond to complex, uncertain environments and multiple users (Chiu, 2005), pure neural network learning systems lack logical reasoning mechanisms. Fuzzy theory seeks to pair the advantages of both approaches, while avoiding their disadvantages.

4.3 Situation simulation

In order to verify the feasibility of applying a neuro-fuzzy approach, this study used the following planning processes as the basis for the design of a learning agent in a simulated situation: (1) Fuzzy logic inferences: When linguistic term descriptions are input, the rule-based fuzzy inference plan gives a degree of support (DoS) initial value (which is usually as 1 to indicate a highly supported rule). Fuzzy inference is then preliminarily used to output the action value (pre-adjustment). (2) User adjustment and records: Output action values are adjusted on the basis of users' actual use (post-adjustment), and the result of adjusting the action of architectural elements is recorded and stored in a database. (3) Neuro-fuzzy training: The database provides examples for neuro-fuzzy network training. Computational learning adjusts the DoS, and training continues in a cyclic fashion until the error between use and the fuzzy logic and neuro-fuzzy system is minimized, at which point training ceases. Alternately, adjustment (post-learning) may stop after the degree of adjustment is less than a certain preset threshold value. (4) When the DoS have been adjusted, the fuzzy logic inference plan will be optimal, and the post-learning output value should be closer to the post-adjustment output value than to the pre-adjustment value (Fig. 12).

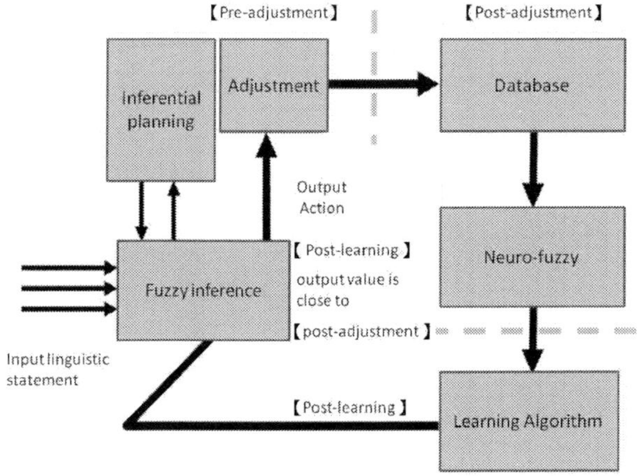

Fig. 12. Planning processes in a simulated situation

4.3.1 Situation simulation

The main task in this simulated situation was the adjustment of indoor lighting, which was performed by different agents. The Fuzzy-TECH software was used to simulate a smart skin's fuzzy logic inferences and neuro-fuzzy learning. The unit modules in this experiment were simplified as two input terminals and one output terminal, and linguistic terms were simplified to three levels (e.g., low, mid, and high).

4.3.2 Setting user attributes and activity types

The agents output adaptive actions with different smart care levels, and the actions can be seen as response functions of user age and activity needs:

IF $<$*user age, activity needs* $>$*THEN* $<$*action* $=$*F(user age, activity needs)*$>$

The goal of setting user attributes and activity type is to test adaptive actions with different smart care levels. In accordance with observations of everyday life, the chief causes of differences in the actions of agents are: (1) User age. As age increases, the user's vision gradually deteriorates, and the user needs more light to support activities (physiological need). (2) Lighting needs of different activities. Different lighting levels are needed for users' different activities (environmental need). (3) Activity privacy needs. Different amounts of spatial privacy are needed to support different activities (psychological need).

The user attribute categories consisted of adults over 30 years of age and seniors under 70 years of age. The 30 users included equal numbers of men and women. In accordance with their user-oriented smart care level, the occupants were classified as normal, special disabled persons, and healthy seniors. The lighting needed for the users' activities was classified as dim (for relaxation—resting, talking), medium (for general tasks—reading, writing), and bright (for precision tasks—sewing, nursing care). In addition, activity privacy needs were classified as low (e.g., talking), medium (e.g., reading, writing, sewing), and high (e.g., resting, nursing care).

4.3.3 Establishment of environmental situation and simulated process framework

This experiment used a window agent as example smart skin, and investigated the possibility of coordination and cooperation between a smart skin and other agents. The experiment was conducted in a 3.6 m x 3.6 m x 3.6 m indoor space. Light was obtained through a south-facing window; the solar altitude was fixed at 45°, and the sky brightness was set at 500 cd/m2 (Fig. 13). The windowsill height was 90 cm above the floor, and the window opening was 2.7 m x1.8 m (w, h). The temporary furniture arrangement consisted of a sofa, a reclining chair, a work table, and chairs, and was intended to facilitate various activities. (Fig. 14).

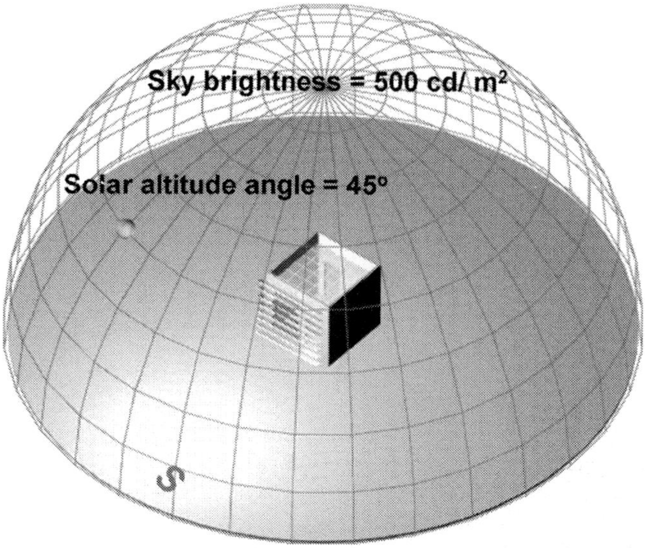

Fig. 13. Lighting environment settings

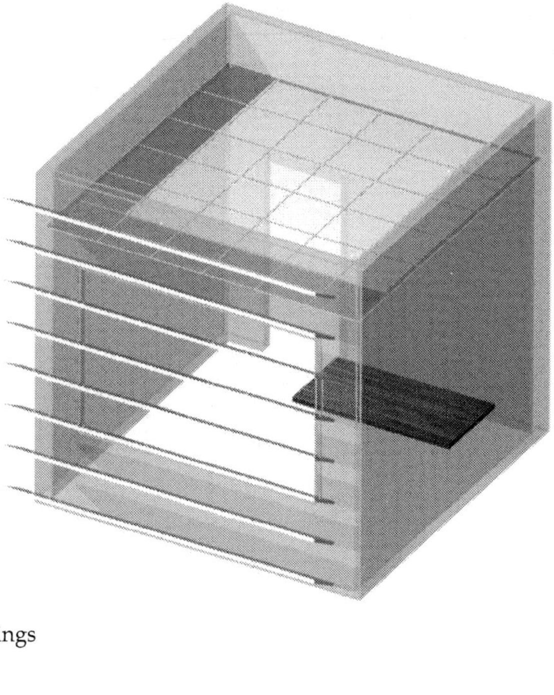

Fig. 14. Spatial settings

The window agent consisted of two subagent module elements: A louver board (LB) agent and a polymer-dispersed liquid crystal (PDLC) glass agent. PDLC glass contains minute liquid crystal droplets dispersed in a polymer grid. An optoelectronic effect allows the transparency of the glass to be changed. The LB agent adjusted the louver angle (down, zero, up) in accordance with indoor activity lighting needs (dim, mid, bright) and the user's view needs (low, mid, high). The PDLC glass agent served to adjust the transparency of the PDLC glass in accordance with the indoor activity lighting needs (dim, mid, bright) and the user's privacy needs (low, mid, high). Furthermore, the system interacted with another smart entity—a lamp agent. The system received information (including louver angle and PDLC glass transparency) from the window agent via wireless signals, and adjusted lamp brightness in order to improve indoor illumination (Fig. 15, Fig. 16).

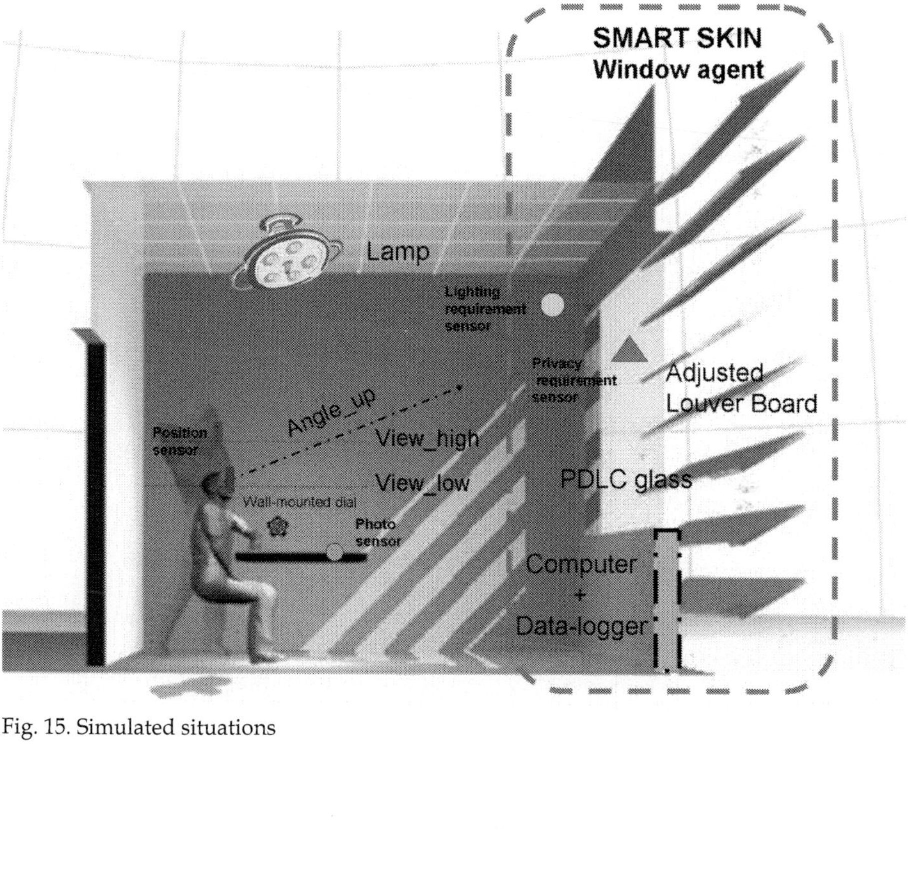

Fig. 15. Simulated situations

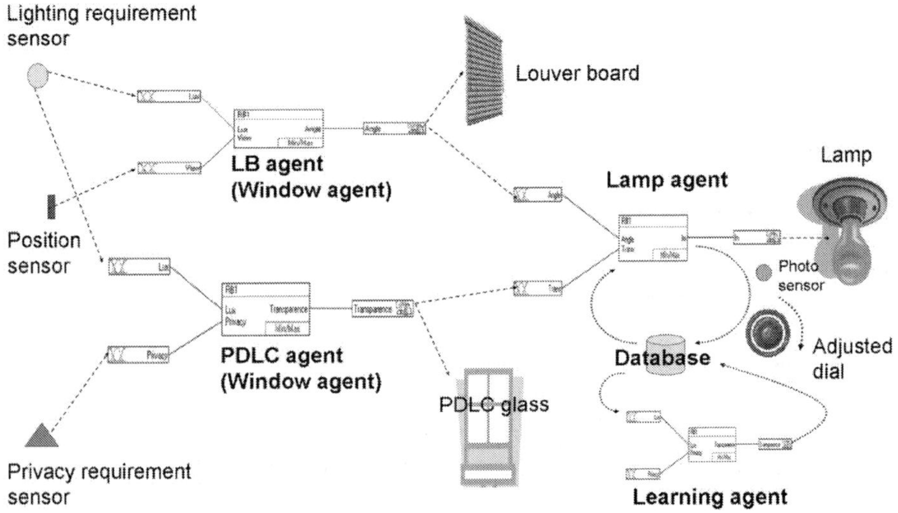

Fig. 16. Interaction and cooperation between agents

According to software simulations, although the missions of the different agents, and the goals and interests of the agent communities, were not necessarily identical, communication, compromise, and conflict involving the causal relationships between perceived events and events needs were handled by means of rational inference and arrangement depending on the fuzzy inference plan. The agents' fuzzy inference plan: IF-THEN rule inferences and matrix rules. As can be seen from Table 2, a matrix rule represents the causal relationship between two perceived events and an output action. For instance, when the LB agent perceives that the light need is high (Lux_bright) and view need is low (View_low), it will adjust the louver angle to Angle_up in order to obtain more sunlight and a better view. When the PDLC agent perceives that the light need is high (Lux_bright) and privacy need is high (Privacy_high) it will reduce the transparency of the PDLC glass (Trans_low).

	Input IF <E(P, E)>	Output THEN <A(B, EA)>	Matrix Rule
Window Agent			
L B Agent	Lux_dim=(1/0, 1/200, 0/500) Lux_mid=(0/200, 1/750, 0/1300) Lux_bri=(0/1000, 1/1500, 1/2000) X View_low=(1/0, 1/90, 0/100) View_mid=(0/90, 1/110, 0/130) View_hig=(0/120, 1/140,1/160)	Angle_Down=(1/-90,1/-45,0/0) Angle_Zero=(0/-5,1/0,0/5) Angle_Up=(0/0, 1/45, 1/90)	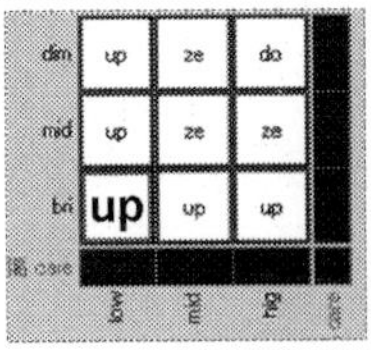
PDLC Agent	Lux_dim=(1/0, 1/200, 0/500) Lux_mid=(0/200, 1/750, 0/1300) Lux_bri=(0/1000, 1/1500, 1/2000) X Privacy_low=(1/0, 1/200,0/300) Privacy_mid=(0/200, 1/300,0/400) Privacy_hig=(0/300, 1/400,1/500)	Trans_low=(1/0, 1/30, 0/50) Trans_mid=(0/30, 1/50, 0/70) Trans_hig=(0/50, 1/70, 1/100)	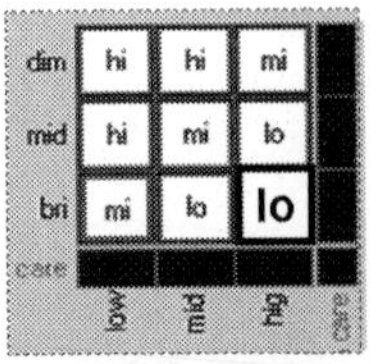
Lamp Agent			
	Angle_Down=(1/-90,1/-45,0/0) Angle_Zero=(0/-5,1/0,0/5) Angle_Up=(0/0, 1/45, 1/90) X Trans_low=(1/0, 1/30, 0/50) Trans_mid=(0/30, 1/50, 0/70) Trans_hig=(0/50, 1/70, 1/100)	Im_low=(1/0,1/200, 0/500) Im_mid=(0/300,1/900, 0/1500) Im_bri=(0/1200,1/1800, 1/2400)	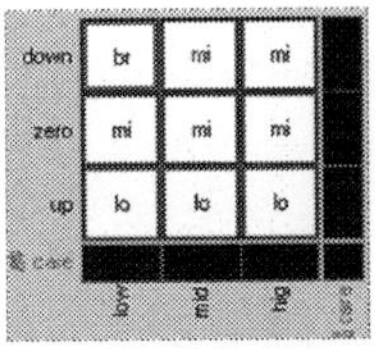
Database Agent			
	Used Record of Lighting	To Learning Agent	
Learning Agent			
	Training set data from Databased Agent	Dos values are adjusted	

Table 2. Fuzzy inference plan

If the indoor brightness is adjusted on the basis of the lamp agent's inferences, but the user is dissatisfied, it is also possible to adjust the lighting using a dial on the wall. As shown in Fig. 16, the simulated situation supports the foregoing agent interactions and users' activities. Relying on a database containing records of lamp use, a learning agent can use neuro-fuzzy computing training data sets to adjust the DoS for the lamp agent's fuzzy inferences, and thereby enhance the lamp agent's ability to predict user behavior. The DoS indicates the user's level of support for or satisfaction for the fuzzy rule in question. The pre-adjustment DoS value is set as 1 as an initial hypothesis. This expresses a high degree of expert support for that rule. The post-adjustment value is the revised value after use (Fig. 17), and expresses the difference between the expert rule and actual use. Fig. 18 records the lamp agent's adjusted DoS values after learning.

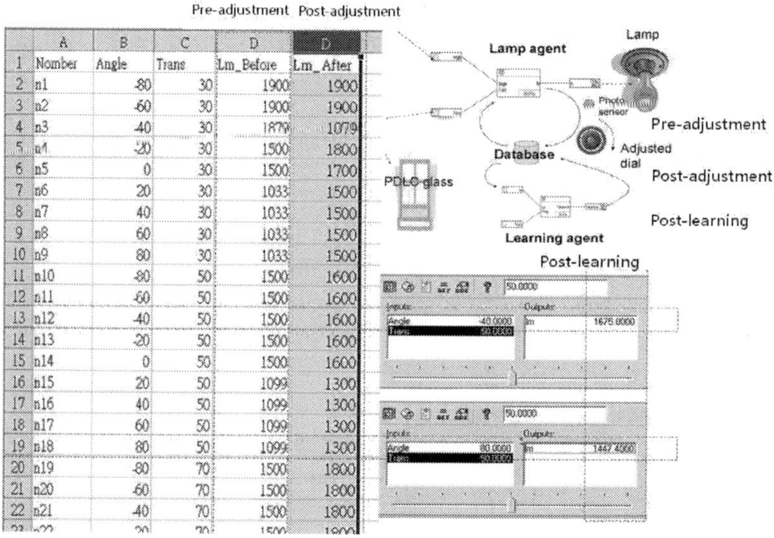

Fig. 17. Recorded pre-adjustment and post-adjustment input and output values; the post-learning values are shown on the right

#	IF Angle	Trans	THEN DoS	lm	#	IF Angle	Trans	THEN DoS	lm
1	down	medium	1.00	mid	1	down	medium	0.53	mid
2	down	medium	1.00	bri	2	down	medium	0.97	bri
3	down	up	1.00	mid	3	down	up	0.61	mid
4	down	up	1.00	bri	4	down	up	1.00	bri
5	zero	low	1.00	mid	5	zero	low	0.55	mid
6	zero	low	1.00	bri	6	zero	low	0.96	bri
7	zero	medium	1.00	mid	7	zero	medium	0.57	mid
8	zero	medium	1.00	bri	8	zero	medium	0.93	bri
9	zero	up	1.00	mid	9	zero	up	0.59	mid
10	zero	up	1.00	bri	10	zero	up	1.00	bri
11	up	low	1.00	low	11	up	low	0.25	low
12	up	low	1.00	mid	12	up	low	0.77	mid
13	up	low	1.00	bri	13	up	low	0.91	bri
14	up	up	1.00	mid	14	up	up	0.87	mid
15	up	up	1.00	bri	15	up	up	1.00	bri
16	down	low	1.00	bri	16	down	low	1.00	bri
17	up	medium	1.00	bri	17	up	medium	1.00	bri
18	up	medium	1.00	low	18	up	medium	0.48	low
19	up	up	1.00	low	19	up	up	0.45	low
20					20				

Fig. 18. Post-learning adjustment of DoS value

5. Conclusions

An agent-based control system can be divided into two parts responsible for describing and setting the responses and actions of intelligent devices. The first of these consists of an independent intelligent agent module and its computing mechanism and plans, and the second consists of the intelligent agent community and its interaction model. Under fuzzy logic operating conditions, each intelligent agent is a clearly defined smart module, and agent communities can rely on cooperation and interaction to achieve their design missions. If a user experience-oriented context awareness function is taken as an agent design goal, fuzzy logic is superior to other inference mechanisms insofar as it can provide a near-human classification of feelings and sensations and inference method, and can also generate continuous mechanical output effects. In other words, such a system does not need to comply with the threshold value restrictions of rule-based reasoning, which cause actions to be discontinuous, nor do binary logic conflicts cause actions to be interrupted. Apart from possessing human or biological learning characteristics and transmitting experience derived from experts' inferences, neuro-fuzzy learning avoids erroneous and defective learning. Situation simulation results displayed that the use of agents possessing only inferential computing ability in the establishment of a user experience-oriented context awareness function is insufficient. Instead, agents must possess learning ability, and be able to rely on constant learning to achieve familiarity with and acquire users' life experience. The experiment confirmed that agents can rely on fuzzy logic inferences and neuro-fuzzy learning to use examples of user experience to adjust degree of support for fuzzy rules, and thereby eliminate the difference between expert rules and actual use.

Furthermore, in an environment containing many complex factors, independent agents cannot easily complete their missions in isolation. Instead, a community of agents must rely on coordination and cooperation to resolve the difficulties that it faces. The complexity of real environments often provides independent agents from using weighting alone to resolve problems. In particular, when agents' actions cause conflicts, and a dilemma occurs, the coordination and cooperation of an agent community are needed to eliminate the problem. In an example earlier in this paper, when loud construction noise from outside can come in through an opening in a classroom wall, and the classroom is extremely hot because of crowded students, would it be better to open or close the window in order to achieve a comfortable classroom environment? As described above, the outdoor noise level can be classified as low, medium, or high, and the indoor temperature can be classified as low, medium, or high. Consequently, IF < room temperature high> THEN < window open>; IF < noise high> THEN < windows closed>. Nevertheless, IF < noise high and room temperature high> THEN < the window should be open or closed?>. At this time, a fuzzy rule can be employed to solve a problem with multi-value inputs and a binary output. Although, in theory, variables and rules can give agents a multi-value model, in the real world there are numerous binary actuators. As a result, even if inference using fuzzy rules can yield multi-value outputs, the restrictions on real actuators may cause fuzzy rule inference problems to revert to binary logic rule problems.

To resolve multi-value input, binary output decision-making problems, (1) transfer functions must be added after fuzzy rule inference. For instance, when an output value has a range of [0,1], the transfer function will be defined as: 0 (Off) when the value is $\leqq$ 0.4,"

XOR" for values between 0.4 and 0.6, where 0.5 is the highest "XOR" point, and 1 (On) when the value is ≥ 0.6. This allows multi-value output values derived by fuzzy rule inference to be converted to three kinds of situations, namely binary output values (On and Off) and "XOR". (2) When "XOR" occurs, the signal must be transmitted to a coordination agent. In an agent community, coordination agents are not necessarily responsible for final decision-making. Instead, they can eliminate the causes of "XOR" states; when the causes are eliminated, decision-making becomes easier. For instance, when a coordination agent has to deal with two variables, such as sound and temperature, regardless of whether the window is opened or closed, or whether it should be partially opened following a weighting process, the agent will face a dilemma. The coordination agent must coordinate with other agents. For instance, if it decides to close the window to shut out outdoor noise, it must cooperate with the air conditioning agent, which must turn on the air conditioning to lower the indoor temperature. Alternatively, if it decides to open the window, it must cooperate with the campus activity management agent, which can re-schedule construction work to non-class hours, and thereby eliminate disturbance from construction noise.

In summary, agent theory emphasizes that intelligence is jointly created by many dispersed, simple independent agents. These agents can fulfill their common missions and goals reflecting complex factors through coordination and cooperation. The application of a neuro-fuzzy computing mechanism can enable agents to achieve cognitive learning and perform adaptive actions. As a consequence, a smart skin possesses a user experience-oriented context awareness function.

Acknowledgements: NSC98-2221-E-035-074, NSC98-2218-E-035-003- grant and Architecture & Building Research Institute, MOI assistance for "981- Smart Living Spaces" class plan.

6. Reference

Altrock, C. V., (1995), (Fuzzy Logic and Neuro Fuzzy Applications Explained), Prentice Hall, 1995, pp. 18~27, 70, 79, 272~277, ISBN-10: 0133684652; ISBN-13: 978-0133684650

Chiu, M. L. And S. Y. Chen, (2005), Exploration of Smart Home Context-Aware Application-Smart-Skin as an example, (Proceedings of The Fifth China Urban Housing Innovation Conference), Hong Kong, PP.81~87, ISBN : 7-112-07847-4

Dey, A.K, (2000), (Providing Architectural Support for Building Context-Aware Applications), Ph.D. Thesis, December 2000, College of Computing, Georgia Institute of Technology

Junestrand, S., Tollmar, K., (1999), Video Mediated Communication for Domestic Environments: Architectural and Technological Design, Streiz, N., J Siegel, V Hartkopf and S Konomi (eds.), (Cooperative Buildings: Integrating Information, Organizations and Architecture, Proceedings of CoBuild'99. LNCS 1670, Springer), pp. 176~89, ISBN-10: 354066596X; ISBN-13: 978-3540665960

Mahdavi, A., (2005), Sentient Buildings: from concept to implementation, in Chiu, M. L. (ed.), 2005, (Insights of Smart Environments), Archidata, pp. 45~66, ISBN : 957-0454-66-0

Minsky, M., (1988), (The society of mind), Simon & Schuster, Inc., pp. 17~37; 34~35; 193, ISBN-10: 0671657135; ISBN-13: 978-0671657130

Mozer, M. C., (1998), The Neural Network House: An Environment that Adapts to its Inhabitants, in M. Coen (ed.), (Proceedings of the American Association for Artificial Intelligence Spring Symposium), Menlo Park, CA: AAAI press., pp. 100~114

Mozer, M. C., (2005), Lesson from an adaptive Home, D. J. Cook and S. K. Das (ed.), (Smart environments), John Wiley & Sons, Inc., pp. 273~294, ISBN : 0-471-54448-5

Negnevitsky, M., (2005), (Artificial Intelligence A Guide to Intelligent System), Second edition published, Person education, Ltd, pp.26~54, 87~129, 268~284, ISBN-10: 0321204662; ISBN-13: 978-0321204660

Padgham, L. And M. Winikoff, (2004), (Developing intelligent agent systems a practical guide), John Wiley & Sons, Ltd, pp. 21~31, 55, ISBN 0-470-86120-7

Russell, S., Norving, P., (2003), (Artificial Intelligence: A modern Approach), (2nd ed.), Pearson Education, Inc., pp. 1~4, 32~58, 194~239, 462~491, 649~677, 764~765, ISBN: 0137903952

Selker, T., W. Burleson, (2000), Context-aware design and interaction in computer systems, (IBM systems Journal, Vol 39, NCS 3&4)

Wooldridge, M. J., (2002), (An Introduction to Multi-Agent Systems), John Wiley & Sons, pp. 1~3, 23~27; 105~106, ISBN-10: 047149691X; ISBN-13: 978-0471496915

5

An AmI-enabled OSGi platform based on socio-semantic technologies*

Ana Fernández Vilas, Rebeca P. Díaz Redondo, José J. Pazos Arias, Manuel Ramos Cabrer, Alberto Gil Solla and Jorge García Duque
Department of Telematics Engineering, University of Vigo
Spain

Abstract
Ideally, smart homes should make its inhabitants' lives more comfortable by anticipating their needs and satisfying their preferences. With this aim, we introduce an approach for context-aware personalized smart homes based on three pillars: the Open Service Gateway Initiative (OSGi) platform, the Semantic Web philosophy and the collaborative tagging trends. By combining these fields, we enrich the OSGi service-oriented architecture by providing a semantical conceptualization of services at home. On the top of this semantic formalization of services, we support context-awareness personalization by using a dynamic and non previously agreed structure for modeling context: a folksonomy. This socio-semantic approach to the problem takes into account the heterogeneous nature of the devices which provide contextual information so that defining a previously agreed constrained vocabulary for context is unrealistic.

1. Introduction

To realize the vision of Ambient Intelligence (AmI), smart homes should make its inhabitants' lives more comfortable by anticipating their needs and satisfying their preferences, that is, smart homes should be able to automatically select and provide the adequate services to its inhabitants' at the right time. This personalization should be done accordingly to the preferences and behavior of the inhabitants and their surrounding environment. Obviously, this goal entails joining efforts coming from different research fields like residential gateways, context-awareness, personalization, sensors, home networks, etc. With this aim, we introduce an approach for context-aware personalized smart homes based on two main pillars: the Open Service Gateway Initiative (OSGi (*OSGi Service Platform, Core Specification, Release 4*, 2005)) platform and the Semantic Web philosophy. By combining both fields, we enrich the OSGi service-oriented architecture by providing a semantical conceptualization of (i) services at home, (ii) contextual information and (iii) inhabitants' preferences. This ontological structure supports reasoning about the captured behavior and inferring new knowledge.

*Work funded by the Ministerio de Educación y Ciencia (Gobierno de España) research project TSI2007-61599, by the Consellería de Educación e Ordenación Universitaria (Xunta de Galicia) incentives file 2007/000016-0, and by the Programa de Promoción Xeral da Investigación de la Consellería de Innovación, Industria e Comercio (Xunta de Galicia) PGIDIT05PXIC32204PN

OSGi specification is currently the most widely adopted technology for building control systems for the networked home. However, its mechanism for service discovering (based on syntactic match making) and invocation assumes the client to know both the name of the service interface and its method's signatures, respectively. Because of this, in previous work we have defined a Semantic OSGi platform (Díaz Redondo et al., 2008) inspired by the Semantic Web conception (Berners-Lee et al., 2001), which is based on the markup of OSGi services by using appropriate ontologies. In the Semantic OSGi platform we have defined, OSGi services describe their properties and capabilities so that any other software element can automatically determine its purpose (semantic discovery) and how to invoke them.

The work introduced in this paper means another step to enrich the Semantic OSGi platform to support both context-awareness and personalization. We pursue a platform which is able to learn about the preferred services for inhabitants in specific contexts and, consequently, to invoke those appropriate services when applicable. This entails the services must be automatically discovered, selected according to both the contextual information and user preferences and launched using the most adequate invocation parameters for the environmental atmosphere at home. In order to do that, we postulate that a previously agreed vocabulary for modelling context does not respond to the dynamic, heterogeneous and distributed nature which is intrinsic to the sources of contextual information. In such an environment, the top-down strategy in ontologies, which remains valid for services categorization, turns into inappropriate. A solution which establishes the context model from the bottom (from the devices, sensors, services or even inhabitants at home in an unconstrained way) can make context management simple, flexible, interoperable and maintenable.

To introduce the context-aware personalized extension to OSGi we propose, the paper is organized as follows. Sect. 2 and Sect. 3 overview the characteristics of the current OSGi framework and the Semantic OSGi platform we have previously defined (Díaz Redondo et al., 2008). Then, Sect. 4 introduce the notion of context and preference in out knowledge base. The details related to our proposal of adding context-awareness and personalization to the Semantic OSGi platform is described in Sect. 5 and Sect. 6. Once the approach has been introduced, the details about the implementation of the prototype are presented in Sect. 7. Finally, Sect. 8 is devoted to analyze other approaches related with the smart home technologies. Finally, a brief discussion including conclusions and future work is presented in Sect. 9, respectively.

2. Overviewing the Semantic OSGi Platform

The OSGi platform consists of a Java Virtual Machine (JVM), a set of running components called *bundles*, and an OSGi framework (*OSGi Service Platform, Core Specification, Release 4*, 2005). A bundle is a Java archive (JAR) and the minimal deliverable application in OSGi, whereas the minimal unit of functionality is a *service*. An OSGi service is defined by its Service Interface, specifying the service's public methods, and is implemented as a Service Object, owned by, and runs within, a bundle. So, a bundle is designed as a set of cooperating services, which any application might discover after the services are published in the OSGi Service Registry.

OSGi aims to provide a flexible platform. However, service discovery and use assume the client has a great deal of information about the service it wants to use. On the one hand, since service discovery is based on syntactic match making, the client must know the service's interface name or at least the keywords in its properties. This mechanism has several drawbacks, including problems with synonyms (semantically similar but syntactically different services) and homonyms (syntactically equivalent but semantically different services). On the other hand, for invocation purposes the client must know the service's method signatures. This is clearly an obstacle in a pervasive environment because it prevents bundles without prior knowledge of the service interface from dynamically invoking the service.

These requirements conflict with the remote management of OSGi applications, one of the platform's mainstays. How can OSGi applications be remotely deployed and work properly on every customer platform if home devices can differ from one customer to another? What is more, as the OSGi specification notes, customers can use the home service platform to run services from different service providers. So, it is hardly realistic to suppose that a specific provider knows, a priori, the interfaces that match other providers' services.

In this context, OSGi's simple syntactic name matching is not enough. So, we proposed a semantic approach to service discovery that turns OSGi into a Semantic OSGi platform (Díaz Redondo et al., 2008). Our approach was based on the markup of OSGi services by using appropriate ontologies, since they support a common understanding for both the service requester and the service provider. More precisely, the table-based structure in the OSGi Service Registry was replaced by an ontological structure integrating: semantic classification, semantic description of properties and information of invocation. Additionally, we have also provided a way for OSGi bundles to register semantically their services and to get the correct references to the needed ones. The work in this proposal is summarized in the following sections.

2.1 The OWL-OS Framework

The main contribution of the OWL-OS framework is re-structuring the table approach in OSGi Registry into an ontological structure which conceptualizes the main aspects of the smart home, that is, its resources and its services. The former can be conceptualized by any structure; OWL-OS is agnostic at this respect. The latter relies on OWL-OSGi Services ontology (OWL-OS) in Fig. 1. OWL-OS semantically describe OSGi services as an extension of OWL-S (*OWL-S: Semantic Markup for Web Services. 1.1 Release*, 2004), and OWL-based Web Services Ontology which provides a semantic markup language specially thought to specify Web Services in unambiguous and computer-interpretable form.

2.2 The OWL-OS Ontology

The OWL-S ontology is organized to provide three essential types of knowledge about any `Service` provided by a `Resource` (upper part of Fig. 1): (i) the `ServiceProfile` tells *"what the service does"*; (ii) the `ServiceModel` explains *"how the service works"*; and (iii) the

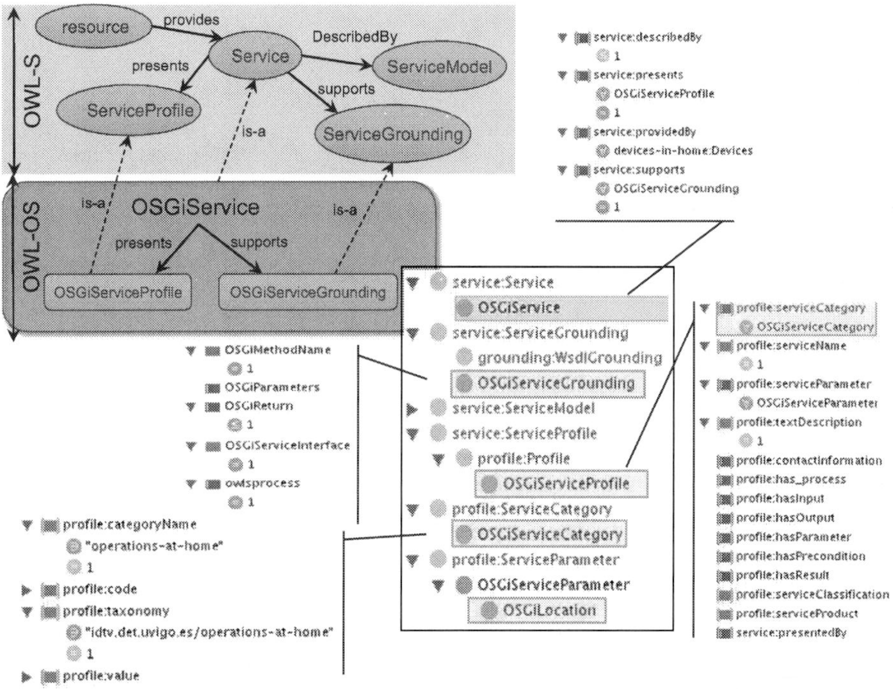

Fig. 1. The *OWL-OS ontology* (OWL-OSGi Services Ontology)

`ServiceGrouding` details *"how to access it"*. We have adapted this ontology to the peculiarities of the OSGi services by defining the OWL-OS ontology (*OWL-OSGi Services* ontology).

So, an *OSGiService* is characterized by: (i) its *OSGiServiceProfile*; (ii) its *OSGiServiceGrounding*; and (iii) some instance in the ontology which supports smart home modelling[1]. Finally, we do not adopt any restriction regarding the *ServiceModel* (how the service works) because the OWL-S one fits perfectly with the OSGi perspective.

Inspired in the OWL-S ServiceProfile, the **OSGiServiceProfile** (subclass) describes an OSGi service by using two main fields (see Fig. 1): (i) the *OSGiServiceCategory* classifies the service in an ontology representing the typical services at the smart home (`operations-at-home` in our case[2]); and (ii) the *OSGiServiceParameter*, an expandable list of properties that may help the requester to find the service. Note that any OSGiServiceProfile may provide more than one categorical description; for instance, the service *"opening a window"* can be both an

[1] As it is said before, the solution is agnostic with respect to the conceptualization of smart home domain. Anyway, we use `devices-in-home` ontology for implementation

[2] Once again the solution is agnostic with respect to this ontology.

airing service or a lighting service. Any other information available about the OSGi service is maintained as a list of OSGiServiceParameters. Finally, the **OSGiServiceGrounding** (subclass of the ServiceGrounding) provides a suitable way for mapping the OSGiServiceProfile to the functionality offered, through a Service Interface, by a bundle running in the OSGi framework.

3. Operation in OWL-OS Framework

The operation in the semantic Framework imitate the way of doing in an OSGi platform. An OSGi service is defined by a *Service Interface*, specifying the service's public methods, and is implemented as a *Service Object*, owned by, and runs within, a bundle. The bundle is in charge of registering the service object with the OSGi Service Registry so that its functionality is available to other bundles. For instance, in Fig. 2 bundle A registers the service object (`serviceobj`) jointly with a dictionary object (`props`), under the `Printing` Service Interface as follows:

```
registerService("Printing", serviceobj, props)
```

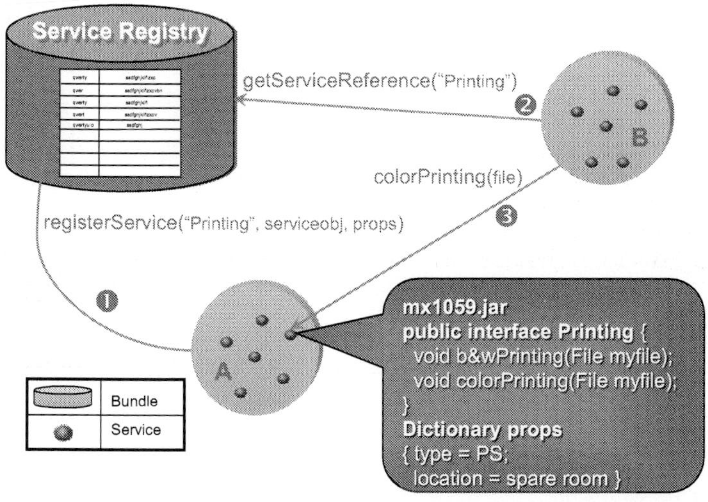

Fig. 2. Operation in the OSGi Platform

Secondly, registered services are referenced through *Service Reference* objects, which maintain the properties and other meta information about the service. Bundles can obtain an OSGi service, actively by syntactic search (`getServiceReference()`), or passively by the event mechanism provided by the framework (*Service Listeners*). For instance, bundle *B* can obtain the Service Reference object corresponding to the service registered by bundle *A* as follows:

```
ref = getServiceReference("Printing")
```

Once bundle *B* has obtained the Service Reference object, it must obtain the service object itself to be able to use the required method, `colorPrinting` in this case:

```
printer = getService(ref)
printer.colorPrinting(file)
```

On another hand, the OSGi framework also offers the possibility of refining the search by using the `getServiceReferences()` method, whose input parameters are (i) the name of the Service Interface and (ii) filtering information to help the matching process. The filter syntax is based on the LDAP language (*Lightweight Directory Access Protocol*) (Howes, 1996). For instance, to obtain the list of all PostScript services belonging to the Service Interface `Printing`, a bundle could use the command:

```
ref=getServiceReferences("Printing", "(type=PS)")
```

where the filter `type=PS` is matched against the dictionary of properties provided on registration.

3.1 Operating semantically

The OWL-OS framework redraw the OSGi operation (registering, discovering and invoking OSGi services) in the following way. For **registering purposes**, we propose the bundle developer specifies the semantic information in the bundle's manifest file using a set of new headers: the *OWL-OS headers*. According to this information, the Semantic OSGi Service Registry may create, if necessary, new individuals and/or classes when a new service is registered in the ontology. So, OSGi bundles use the same registering method (`registerService()`) of the OSGi specification (see Fig. 3). Note we maintain the dictionary of properties in the registering method just for compatibility reasons with the standard Service Registry. The OWL-OS parameters of service, like `OSGiLocation` or other service-dependent parameters (`printFormat` in our example) are assumed to be managed inside the OWL-OS ontology. Their initial values, if any, are extracted from the manifest file (for instance the value `#PS` for `printFormat`).
For the semantic **discovery mechanism** we have proposed to add a method inspired in the OSGi original one:

```
semGetServiceReference(String OntURI, String Cat)
```

Thus, instead of providing the name of the Service Interface, the requesting bundle specifies the name of the service category (`Cat`) and which ontology is using to classify them (`OntURI`).
Finally, to **invoke** semantic services, we have added get-set style methods to the *ServiceReference* interface of the OSGi Specification. So, the requester bundle can discover the name of the public method and the parameters in order to construct the invoking primitive.

4. Modeling context and preferences

To establish a context-aware and personalized solution on top of our OWL-OS framework, we need an automated model of both the context and the preferences of the user. These two models are not independent each other so that the preferences strongly depend on the context of the user (where, when, etc.).

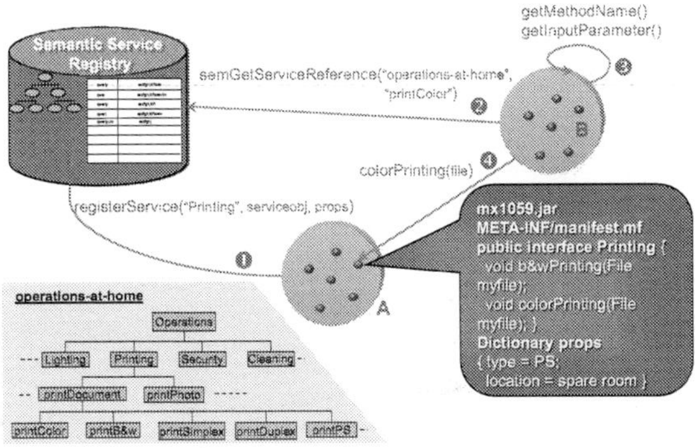

Fig. 3. Operation in the Semantic OSGi Platform

4.1 Context modeling

For context modeling, we assume the Zimmermann (Zimmermann et al., 2005) vision of **CXMS (ConteXt Management System)** as a system which involves the construction, integration, and administration of context-aware behavior. Our proposal assumes an external CXMS that is integrated in the OSGi Framework as one more OSGi service. The key question in the smart home is what structures we use to model the context.

A user's context can be defined broadly as the circumstances or situations in which a computing task takes place (Meyer & Rakotonirainy, 2003). More in general, context includes any information that can be used to characterize the situation of an entity (person, physical or computational object) in any domain (home domain, health domain, etc.). In the smart home domain, the home is plenty of sensors which determine various types of contexts of its inhabitants (location, mood, etc.) while applications in the home use this information to provide context-aware services. The great variety of that information makes the task of formalizing a context model specially complex.

Several works in the literature have tackled with the problem of constructing an ontology for contextual information (see Sect. 8). According to the Gruber's popular definition, an ontology is a description of the concepts and relationships that can exist for an agent or a community of agents. These agents (who offer software/data/services) identify and specify some common conceptualization of the data so that systems can be built which interoperate on those specifications. We believe constructing this conceptualization in a up-bottom and static way is never suitable in a domain which is inherently heterogeneous an even subjective. Instead of formalizing this knowledge by using an ontology, we propose using a folksonomy to model context. The term folksonomy was coined by Thomas Vander Wal (Wal, 2007) as 'the user-created bottom-up categorical structure development with an emergent thesaurus'. This emergent structure comes up from the sensor's and device's metadata in a natural way. Moreover, a folksonomy-based approach to context modeling can aggregate not only real data and metadata from devices at home but even user-provided information about his context. Similar approach has been considered in (Mizzaro et al., 2009) in the field of information retrieval.

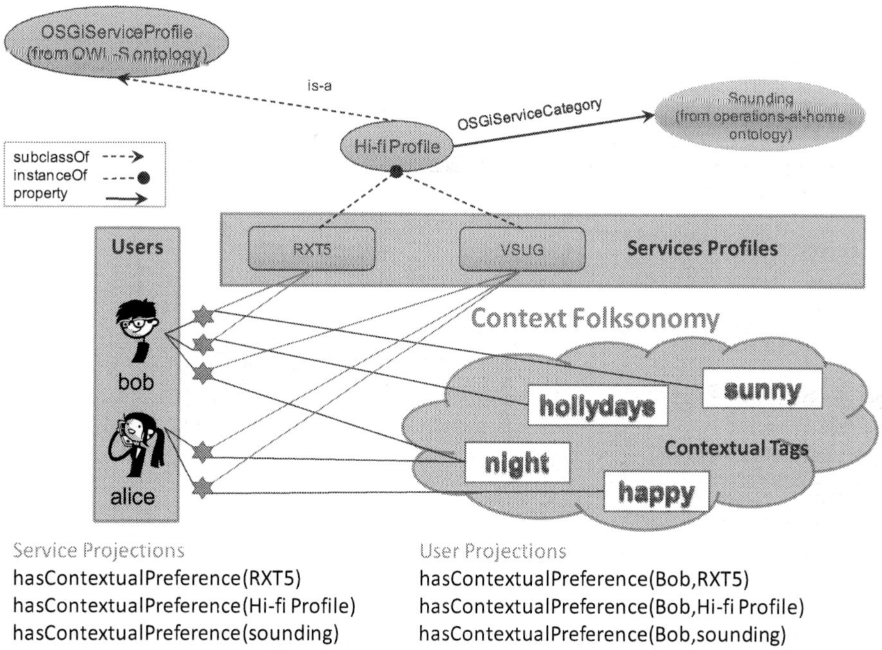

Fig. 4. A folksonomy-based CSMX

In our approach, the actual context of the smart home is represented by the set of tags obtained from the devices and even the inhabitants at home; these tags range over an uncontrolled vocabulary. Beyond the actual context, the information about the contexts in which inhabitants use the services at home is modeled by means of the folksonomic-structure in figure 4. In this figure, we can observe the three elements which are involved in establishing the contextual characterization of a service from the point of view of a specific user (users, service profiles and contextual tags). Mathematically speaking, this folksonomic structure can be described (similar to the characterization in (Hotho, Jaschke, Schmitz & Stumme, 2006b)) as a tuple $F_C := (U, T, S, Y)$ where $U = \{u_1, \cdots, u_m\}$, $T = \{t_1, \cdots, t_l\}$, and $S = \{s_1, \cdots, s_k\}$, are finite sets of items (services in our case), tags (contextual tags in our case) and users. $Y \subseteq UxTxS$ is a ternary relation whose elements are called tag assignments. In our contextual folksonomy for services, tags assignments are the tags which describe the context where an inhabitant uses a service at home. As it is shown in the following section, according to this structure the contextual characterization of a service is represented by a set of tags (graphically by a tag cloud) with their respective weights, which are proportional to the number of users who have used this service in a context described with this tag. Similarly, tags are linked in a way that the more times the tags are assigned together to the same context or situation, the highest the weight of their relationship. This structure is then used to establish relationships between contexts in our semantic OSGi platform.

More than the universe of tag assignments to context, for personalization purposes we want to know the contexts of a specific user. From the tripartite graph above, we can obtain the bipartite graphs of personal folksonomies (personomies), which represents the contextual information of a single inhabitant u using the services at home. Mathematically speaking, the context personomy P_C^u of a user u is the restriction of the folksonomy F_C to u, that is the set of user's tags which are assigned to the set of user's services.

4.2 Preference Model

Until now, we have defined a model to represent the contexts in which users invocate services at home. However, a personalized context-aware solution requires to establish a preference model, also connected with the CXMS, to adapt the context-aware response of the smart home to the user's characteristics. With this aim, we define a **preference model** that stores information about any entity s in the service model (class or instance) from the point of view of any user u in the system, concretely, the context of application C (time, place, etc.) and the preference (an index DOP (*Degree of Preference*)) about the invocation of this service by this user. The value of the index $DOP(u,s)$ reflects how much the user u is interested in using the service s; and the element $C(u,s)$ reflects the characteristics of the past contexts where the user has accessed to the service s.

Based on the above conception, our preference model (Fig. 5) enriches the OWL-OS ontology with the property `hasContextualPreference`; a complex property which consists of two sub-properties: `hasContext` and `hasDOP`. The ontological preference model of the inhabitant u is a projection of this ontology where each entity s (instance or type, say class, of service) is characterized by the contextual preference of the specific user, that is, the context $C(s,u)$ where the service has been used by this user, and the index of preference $DOP(s,u)$ the user about this service. Both properties are defined on top on the ontology so that the pair $\{DOP, C\}$ qualifies every class or instance in the ontology. The value of $DOP(s,u)$ and $C(s,u)$ for a specific entity s in the ontological preference model is recomputed whenever the inhabitant u uses the service by (1) weighting DOP with the number of service instances in the same category and (2) aggregating in C the new contextual information (as it is shown in the following section). From instance-scope DOPs and Cs (`RXT5` and `VSUG` in figure 4) the ones corresponding to classes (`Hi-fi Profile` in figure 4) are obtained by recursively propagating DOPs and Cs up through the hierarchy defined by *operations at home*.

The preference model allows to vary the quantity of collective intelligence to be taken into account when personalizing service selection in a context aware way. At the maximal level of collective intelligence, we have a projection of the ontology over the contextual folksonomy. In that case the `hasContextualPreference` property ranges over the contextual folksonomy F_C and the DOP is computed for all the users considered in the system. At the minimal level

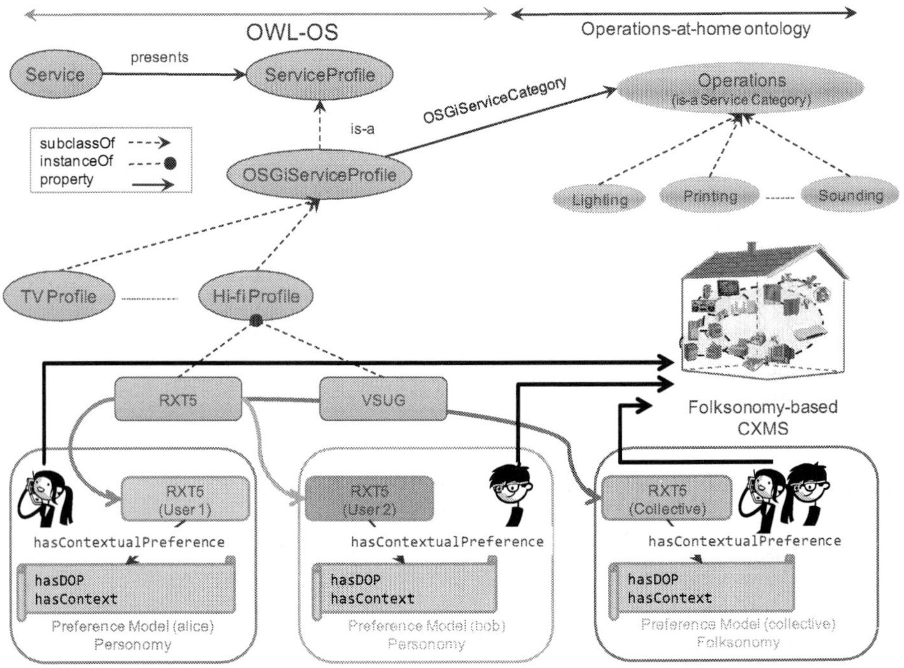

Fig. 5. Context-aware Preference Model for the services at home

of collective intelligence, we have a projection of the ontology over a contextual personomy. In that case the `hasContextualPreference` property ranges over the contextual folksonomy F_C^u and the *DOP* is computed only for the specific user u.

5. Folksonomic support to Context-aware Personalization

5.1 Describing the context of a service

A service in the smart home is contextually described by a tag cloud. From the definition of the folksonomy $F_C := (U, T, S, Y)$, a tag cloud for a specific service profile instance s (say service), is defined as $TC_S(s) = \{t, w(s,t)\}$ such that $(u, t, s) \in Y$ and the weight is computed as follows. For a service s, the weight assigned to a contextual tag t corresponds to the occurrence frequency of the tag in contexts for the service s. We normalize the weights so that the sum of the weights assigned to the tags in the tag cloud is equal to 1.

Being $T(s) = \{t_1, t_2, \ldots, t_{n_i}\}$ the set of tags which have been used to describe the usage context of the service s and $m : T(s) \rightarrow \mathbb{N}$ the function that relates each tag with its multiplicity in the multiset, i.e., the number of times that the tag t have been used to describe the context of the service s, denoted as $m(t, T(s))$. We define the function $w(s,t) : T(s) \rightarrow \mathbb{R}$ as the one which relates each tag $t \in T(s)$ with its weight in the multiset which is calculated as follows:

$$w(s,t) = \frac{m(t,T(s))}{|T(s)|} \tag{1}$$

where $|T(s)|$ is the cardinality of the multiset, i.e., the result of adding the multiplicities of all of its tags:

$$|T(s)| = \sum_{\forall k / t_k \in T(s)} m(t_k, T(s))$$

Hence,

$$\sum_{\forall k / t_k \in T(s)} w(s,t_k) = 1 \tag{2}$$

In this approach, apart from storing the tag cloud for the contextual information of a service, it is also important to know another important parameter, the Index Of Popularity (IOP), calculated from the number of contextual tags assigned to this service with respect to the total number of contextual tags assigned in the system:

$$IOP(s) = \frac{|T(s)|}{\sum_{\forall k} |T(s_k)|} \tag{3}$$

In order to both contextualize and personalize the selection of a service, apart from the tag cloud of a service in the system (the one which characterize the context of the service globally), we also need to know the contextual tag cloud of a service form the point of view of a specific user, that is, the contexts where the user invokes the service. For that, we define the personal $TC_S(s,u) = \{t, w(s,t)\}$ such that $(u,t,s) \in Y$ and the weight is computed as described above. Finally the tag cloud of the actual context is denoted by $TC(home)$ and obtained from the data and metadata provided for all the devices at home, the tags in the tag cloud $TC(home)$ are weighted accordingly.

5.2 Creating a weighted graph for contextualization

In order to properly compare the tag clouds of the actual context at home $TC(home)$ and the one which records the context where a service has been previously accessed ($TC_S(s)$ or $TC_S(s,u)$), we define a weighted graph TG_C for the folksonomy F_C (respectively TG_C^u for the personomy P_C^u). This weighted graph is created from the service contextual tag clouds $TC_s(s)$ in the system (respectively $TC_S(s,u)$ for a specific user u).

TG_C is an undirected graph where the nodes are the tags ever used in the system and the arcs are the relationships between the tags they link. The relationships are calculated from the number of times the contextual tags appear together in a contextual tag cloud, the weights of the tags in this tag cloud, as well as the index of popularity (IOP) of those services described by both contextual tags at the same time. We define the relationship between two tags (t_i and t_j) with respect to one single service s (denoted as $r_{ij}(s)$), as the geometric means between the weights of both tags in $T(s)$.

$$r_{ij}(s) = \sqrt{w(s,t_i)\, w(s,t_j)} \tag{4}$$

To obtain the total relationship between t_i and t_j (r_{ij}), we should take into account their partial relationships with respect to each service in the smart home and appropriately weight them according to its *IOP*.

$$r_{ij} = \frac{\sum\limits_{\forall s / t_i \vee t_j \in T(s)} IOP(s)\, r_{ij}(s)}{\sum\limits_{\forall s / t_i \vee t_j \in T(s)} IOP(s)} \tag{5}$$

5.3 Tag cloud comparison

The selection process need an algorithm of tag cloud comparison. Concretely, the contextual tag cloud which describes the actual context $TC(home)$ has to be compared with the information in the contextual folksonomy in order to decide which service of a specific category is more suitable. As it was mentioned before, in this decision we can vary the quantity of collective intelligence taken into account, that is, using only the information in the user personomy or using the information in the system folksonomy. In this section we propose a metric for comparing contextual tag clouds.

The simplest way to measure this value would be to count the number of coincident tags in both contextual tag clouds, i.e., the higher the number of coincident tags, the higher the degree of relationship between the actual context at home and the contexts where a specific service has been used. Besides, to take into account the relative importance of the coincident tag in the tag cloud, instead of adding 1 for each coincident tag, it is better to add the means of their weights in both tag clouds. The relationships is then calculated like follows:

$$R_0(TC_S(s), TC(home)) = \sum_{\forall i / t_i \in |T(s) \cap T(home)|} \sqrt{w(s, t_i)\, w(home, t_i)} \tag{6}$$

But this approach does not take into account the relationships between the tags in the weighted graph of the folksonomy TG_C. Although a tag that belongs to the first tag cloud do not appear in the second tag cloud (or appears with a lower weight), it can be closely related to the rest of the tags of the second tag cloud. To take this fact into account, we do not only consider the number of coincident tags and their weights (*direct relationship*, $R_0(TC_S(s), TC(home))$), but also the degree of relationship between the tags of both tag clouds (*one-hop relationship*, $R_1(TC_S(s), TC(home))$). In this manner, the total relationship is calculated as follows:

$$R(T_S(s), TC(home)) = \alpha\, R_0(T_S(s), TC(home)) + (1 - \alpha)\, R_1(T_S(s), TC(home)) \tag{7}$$

where the parameter $\alpha \in [0, 1]$ is used to assign more or less importance to the direct or one-hop relationships. The one-hop relationship is calculated by adding the degree of relationship of each pair of possible tags (each tag of the pair belongs to one of the tag clouds) adjusted by the means of their weight.

$$R_1(T_S(s), TC(home)) = \sum_{\substack{\forall i / t_i \in TC_S(s) \\ \forall j / t_j \in TC_(home)}} \sqrt{w(s, t_i)\, w(home, t_j)} \quad r_{ij} \tag{8}$$

Finally, if we want to compare the context at home $TC(home)$ with the context tag cloud of a service from the point of view of a specific user $TC_S(s,u)$, that is in the personomy, the relationships are calculated in a similar way but taking into account the relationships in the weighted graph TG_C^u.

6. Two forms of personalization: the service and its invocation

Although the above preference model allows personalization strategies for devices, providers, etc., we focus on the two most common applications: personalizing the selection of the service and its invocation personalization. Our proposal to provide a context-aware personalization for OSGi services is based on the OWL-OS framework. It defines two logic layers to automatically select (according to the user's preferences and the contextual information) which is the most adequate service and how must it be invoked, respectively. The first stage decides the service in a specific category (*OSGiServiceProfile* entity in OWL-OS); for instance, the most suitable service for playing music. The second one decides how to invoke (parametrisation) the specific service (*OSGiServiceInvocation*); for instance, the most suitable volume when playing music (see Fig. 6).

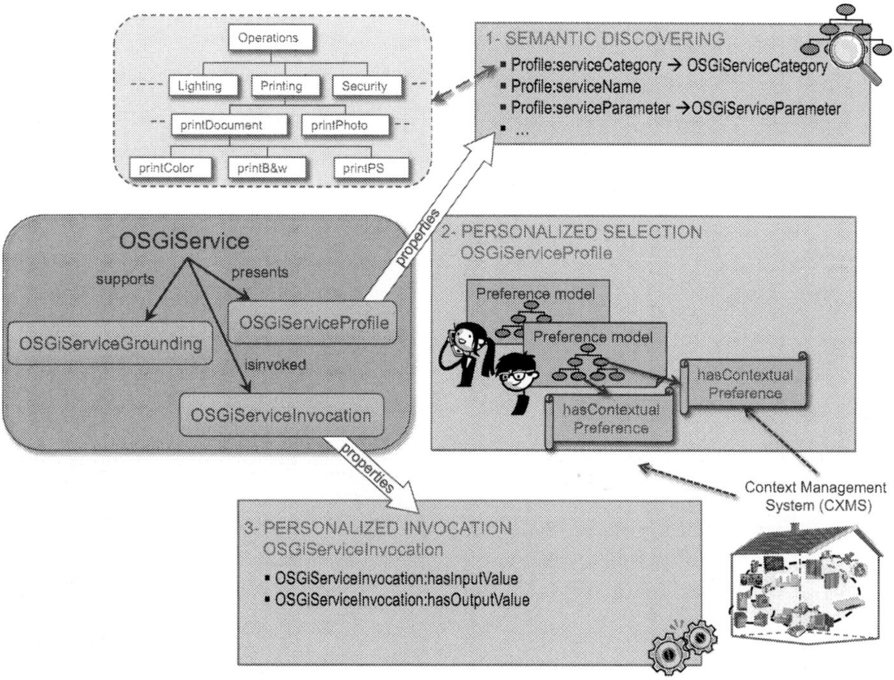

Fig. 6. Using Semantic OSGi for personalizing OSGi services in a context-aware solution

6.1 Context-aware and personalized selection

Service selection can be personalized by using the information in the preference model to establishes the most appealing service for the user U among those which meet the discovering criteria (*OSGiServiceCategory* and *OSGiServiceParameter*) in the current context C. With this aim, we have added another method inspired in the OSGi original ones:

```
semGetServiceReferences(String OntURI, String Cat, String semFilter)
```

This method includes a filtering criterion (`semanticFilter`) based on the SWRL Query instead of LDAP. Thus, the *semFilter* is an OWL query about individuals pertaining to the category specified in the method's second parameter.

From the client's perspective, the same method (`semGetServiceReference` or `semGetServiceReferences`) is used. From the Framework's perspective, the service selection process is quite different and proceeds according to the following steps:

[1] Semantic matching for the service s**:** First, the semantic OSGi Registry looks for those services that match the category (`Cat`) and satisfy the (`semFilter`), if applicable.

[2] Context-aware matching for the user u **in the context** $TC(home)$**:** Now, it starts the personalization part of the service selection using the preference model of the active user u and taking into account the context $TC(home)$. For each service s from step 1, the CXMS receives the `hasContextualPreference` properties in the ontology to do the following steps:

[2.1] Contextual Preference for the user: An index of contextual preference for the user u ($\mathcal{I}_{CP}(s,u)$) is calculated which reflects the preference of the user u about the service s but adjusted by the similarity between the actual context at home and the previous contexts in which the service s has been invoked by the user u:

$$\mathcal{I}_{CP}(s,u) = DOP(s,u)\ R(T_S(s,u), TC(home)) \tag{9}$$

The above parameters are registered in the properties `hasDOP` and `hasContext`) of the service profile instance s for the user u.

[2.2] Contextual Preference for the community: An index of contextual preference for the community ($\mathcal{I}_{CP}(s)$) is calculated which reflects the preference of the community about the service s but adjusted by the similarity between the actual context at home and the previous contexts in which the service s has been invoked by the community:

$$\mathcal{I}_{CP}(s) = DOP(s,u)\ R(T_S(s), TC(home)) \tag{10}$$

The above parameters are registered in the properties `hasDOP` and `hasContext`) of the service profile instance s for the user community.

[3] Context-aware an personalized selection: The semantic OSGi Registry selects the most adequate service, from the ones filtered in step 1, choosing the service s having the highest contextual preference. For this a global index of contextual preference $\mathbf{I}_{CP}$ is defined. This new index allows to select the quantity of collective intelligence to be incorporated in the selection process by adjusting the β parameter in the following equation:

$$\mathbf{I}_{CP}(s) = \beta\, \mathcal{I}_{CP}(s,u) \;+\; (1-\beta)\, \mathcal{I}_{CP}(s) \tag{11}$$

6.2 Context-aware and personalized invocation

One step beyond personalizing service selection is adapting the invocation of that service (typically the values for input parameters) according to the user, his/her habits and the context. Thus, it is possible to capture behaviors like the following: when playing music (the service) John (the user) selects, during the day (context), 90% of the times (*DOP*) high volume; and, at night (context), 80% of the times (*DOP*) low volume. With this aim, and as there is no notion of service execution on OWL-S nor OWL-OS, we introduce the class `OSGiServiceInvocation` in OWL-OS. As Fig. 6 shows, the main properties in this class are: (i) `hasInputValue` and `hasOutputValue`, instances of `Binding` OWL-S classes[3]; and (ii) `hasContextualPreference`.

In order to allow the implementation of the behavior above, we have added specific methods to the OSGi framework: `getInputValue` and `getOutputValue`. The former returns, for a specific service, the preferred value (greater index of contextual preference $\mathbf{I}_{CP}$) among those *OSGiServiceInvocation* entities in the preference model of the user u or of the community depending on the similarity with the current context of the smart home. The latter has the same behavior but referred to an output value. Despite the fact that the application of the preferred output value remains open, it can support, for instance, an automated alarm system which notifies in case of output values different from the preferred ones.

7. Implementation

The prototype of this proposal has been developed over OSCAR (Open Source Container ARchitecture, `http://oscar.objectweb.org`), an open software implementation of the OSGi framework. We have added a semantic version of the OSCAR registry which interprets the new bundle manifests and manages the OWL-OS ontology to search (querying the ontology) and to register services (populating the ontology). Besides, the semantic version of the OSCAR registry maintains the preference models for the house's inhabitants, so it also modifies the services' properties for each profile, like the DOP or the context. To manage the local OWL-OS ontology, the profiles and provide the semantic OSGi services, we use the Protégé OWL API (`http://protege.stanford.edu/plugins/owl/api/`). This open-source Java library provides classes and methods to load and save OWL files, to query and manipulate OWL data models, and to perform reasoning. The CXMS we have designed has been integrated as an OSGi service in the OWL-OS platform. This CXMS manages the context folksonomy and the metrics presented in this paper for modelling and comparing context information.

[3] A binding object has two properties the name of the variable, `toVar`, and the specification of a value for which we suggest `ValueData`.

8. Related Work

Building smart homes entails integrating different computing technologies, such as ubiquitous computing, context-aware computing or home automation technology. This combination supports automatic interactions among residents, computer-equipped devices and the home environment.

Context Modeling entails to represent "any information that can be used to characterize the situation of entities (i.e. whether a person, place or object) that are considered relevant to the interaction between a user and an application, including the user and the application themselves" (Dey, 2001). There are different approaches to organize context knowledge (Strang & Linnhoff-Popien, 2004): from the simplest technique based on a list of attributes in a *key-value* manner to the more sophisticated *ontology-based models*, where ontological structures provide an uniform way for specifying the model's core concepts as well as enabling contextual knowledge sharing and reuse in an ubiquitous computing system; without forgetting graphical, object-oriented or logic-based models. However, having a successful context modeling usually requires of combined strategies. Specially relevant for the smart home, the authors' approach in (Oh et al., 2006) consists of defining any context at home by five concepts (4W1H) : who (resident identification), what (identifies the event that should happens), where (resident's location), when (occurring time) and how (the way the event should develop within these conditions).

OSGi and the smart home. Previous approaches have tried to promote smart spaces by using OSGi, like in (Lee et al., 2003), whose authors propose using OSGi as a suitable infrastructure to integrate various devices and sensors to provide pervasive computing environments. However, they do not resolve the problem of service search and invocation. In (Dobrev et al., 2002) these two problems are directly tackled but not from a semantic perspective. On another hand, the authors of (Gu et al., 2004) propose to define a semantic environment for describing contextual information to be used by context-aware applications. However, OSGi is only used as a support layer, without improving the OSGi framework at all. After an exhaustive revision of the state-of-the-art, we have not found any proposal about integrating semantic reasoning nor personalization within the OSGi framework.

Ontological descriptions for the smart home. Some initiatives have came up in this area. UbisWorld (Heckmann, 2003) is used to represent parts of the real world (a house, a shop, etc.) and all the individuals and elements belonging to it. Its knowledge is modeled by two ontologies which are under development: GUMO (General User Model Ontology) and UbisOntology, the ontology for ubiquitous computing. SOUPA (Standard Ontology for Ubiquitous and Pervasive Applications) (Chen et al., 2004) is another relevant approach. It consists of two set of ontologies: SOUPA *Core* to define generic vocabularies; and SOUPA *Extension* to define additional vocabularies for specific applications. However, no focus has been on the semantic description of the in-home services. For this, we propose to take advantage of the previous efforts made in the Semantic Web field, where the appropriate semantic description of services is the main point. In this area, the most salient initiatives to describe Semantic Web Services are WSMO (Web Service Modeling Ontology) (*D2v1.2 Web Service Modeling Ontology*, 2005) and OWL-S (*OWL-S: Semantic Markup for Web Services. 1.1 Release*, 2004). Although both approaches cover the same field with the same goals trusting ontologies to get them, OWL-S[4] is considered to be clearly more mature in certain aspects and expressive enough to be applied in the smart home environment.

Collaborative tagging and folksonomies Collaborative or social tagging allows users to tag contents and share them in such a way that they can categorize not only content that they themselves have added, but also content added by other users Golder & Huberman (2006). Sites commonly cited as examples of collaborative tagging are Del.icio.us , for the tagging of Web pages, or Flickr, for the tagging of photos. In order to define the structures generated by this type of classification, in 2004 Thomas Vander Wal coined the term folksonomy citepdef-folksonomy. At the same time he established two different types of folksonomies: broad folksonomies, constructed as a result of tagging contents in systems in which any user may tag all the content in the system (as in Del.icio.us), while narrow folksonomies are the result of tagging systems in which only a small number of users may tag content (for example, Flickr, where only the author of the photos and users designated as friends of the author may assign tags to them).

Folksonomies, whether wide or narrow, are structures that can be represented as an undirected graph in which the nodes are the various tags assigned in the system and the transitions are the relationships between two nodes that are joined. These relationships are assigned a weighting which will depend on the number of times tags describing a content item appear together: the higher the frequency, the higher the weighting Michlmayr et al. (2007). A structure of this type lends itself to be used to establish relationships between elements of the system. Some works in the literature have explored a folksonomy-based solution for personalization or recommendation in several domains. The proposal in Niwa et al. (2006) makes use of users' favourite Web pages (obtained from a collaborative tagging system for Web bookmarks such as Del.icio.us) and their tags to recommend new Web pages. To do this it calculates the affinity between users and Web pages, grouping related tags together and finding out which Web pages are the most appropriate to each group. Next it calculates the affinity of the user with the tags of the groups to determine which Web pages are the most appropriate to him or her. Folksonomies can also be used to calculate the relevance of content shown in information retrieval systems; for example, the approach set out in Hotho, Jäschke, Schmitz & Stumme (2006a)Hotho, Jäschke, Schmitz & Stumme (2006) describes an algorithm known as FolkRank (based on the idea of Google's Page-Rank), whereby a resource tagged as important by important users is considered to be important. The principle is applied to both users and tags. Specia & Motta (2007) presents another work in which tags are grouped together according to the relationship between one another. In this case, in order to establish similarity between tags, the system takes into account not only the number of times that two tags appear together but also the number of times that the first of the two tags is used together with tags that, in turn, are used together with the second of the two. Finally, these groups are refined by searching for each pair of tags on a number of different semantic search engines to confirm the relationship between them.

[4] The OWL-S coalition is currently working in an upcoming release 1.2 of OWL-S, whose details are available at `http://www.ai.sri.com/daml/services/owl-s/1.2`.

9. Conclusions and future work

The primary reference architecture in the OSGi specification is based on a model where a operator manages a potentially large network of service platforms. It assumes that the service platforms are fully controlled by the operator and used to run services from many different services providers. In this scenario, we have shown that the actual service discovery mechanisms in OSGi is insufficient. In the pursuit of a really open and interoperable residential gateway, we propose the semantic description and discovery of the services in the OSGi domain. At this respect, we have defined OWL-OS, a sub-ontology of OWL-S which allows making a simple semantic search of services based on a categorized structure. Despite we propose *operations-at-home* as the primary structure to classify the OSGi services, OWL-OS allows an OSGi service to be semantically described according to different ontological structures. These ontological structures would be downloaded on demand from the service provider. Finally, note the Semantic OSGi Framework enhance the OSGi standard, without breaking it; i.e. any non-semantic bundle can work properly within this framework, although it is not able to take advantage of the semantic reasoning for service obtaining. Moreover a clear benefit of the new Semantic OSGi platform is the possibility of supporting the automation of OSGi services composition (Díaz Redondo et al., 2007).

In the field of service selection, it would be possible to supplement the Semantic Registry with different specialized software agents which takes into account other factors (ambient intelligence) to automate the service selection. In this paper we have investigated the potential of combining a personalization agent and a context management system. On its own, the personalization agent contribute to decisions on service selection by using a preference model storing historic information about preferred devices and services. On the other hand, the context management system allows the OWL-OS Framework to be aware of the particular context (time, location, mood, etc) and to react by discovering services which meet that context. Beyond their independent behavior, connecting the personalization and context-aware agent across an ontological base, OWL-OS in our case, allow the OWL-OS framework to support scenarios as the one included in this paper. In general, scenarios where the selection of the service depends not only on who uses the service but when, where or even why uses the service. Although this paper only discusses two forms of context-aware personalization in the smart home (service selection and parametrisation), some other forms are possible with the general preference model we propose. For instance, provider selection can be personalized by capturing the preferences of the user.

Apart from the general approach, one of the main contributions of this work is the effort to provide a flexible context model which fit in well with an extremely heterogeneous environment. With this premise, we provide a collaborative approach to context, in fact we can say that the solution is collaborative in two senses. Firstly, it is a collaborative model of context since all the devices, services, sensors, inhabitants, etc. at home form a pluriarchy without any controlled vocabulary. Secondly, it is also a collaborative model for preferences since the service selection takes into account not only the preferences of the user but even the preferences of a smart home community, possible a community established by the service provider. At this respect, in this paper we have explored algorithms of service selection which rely on tag cloud comparison of contexts, but it is also possible to inspect the similarity between users by comparing user's preference model. Considering the similarity among users, according to their behavior at home, allows the selection process to incorporate more sophisticated strategies of collaborative filtering.

Also as part of our future research, we are working on supporting multi-user preference model. The human presence simulator in this paper is already a multi-user system whose aim is simulating the behaviors of all the inhabitants living in the smart home. The solution to this problem is far from being trivial, because an approach which simply merge the habits of all the inhabitants obviates the relationships and even dependencies among users at home. An adequate multi-user personalization system has to distinguish among the user preferences and cope with conflict resolution.

10. References

Berners-Lee, T., Hendler, J. & Lassila, O. (2001). The Semantic Web, *Scientific American* pp. 35–43.

Chen, H., Perich, F., Finin, T. & Joshi, A. (2004). SOUPA: Standard Ontology for Ubiquitous and Pervasive Applications, *Int. Conf. on Mobile and Ubiquitous Systems: Networking and Services*.

D2v1.2 Web Service Modeling Ontology (2005). The WSMO group.
URL: *http://www.wsmo.org/TR/d2/v1.2/*

Dey, A. K. (2001). Understanding and using context, *Personal and Ubiquitous Computing, Special issue on Situated Interaction and Ubiquitous Computing* **5**(1).

Díaz Redondo, R. P., Fernández Vilas, A., Cabrer, M. R., Pazos Arias, J. J. & Rey López, M. (2007). Enhancing Residential Gateways: OSGi Services Composition, *IEEE Transactions on Consumer Electronics* **53**(1): 87–95.

Díaz Redondo, R. P., Fernández Vilas, A., Ramos Cabrer, M., Pazos Arias, J. J., García Duque, J. & Gil Solla, A. (2008). Enhancing Residential Gateways: a Semantic OSGi Platform, *IEEE Intelligent Systems Journal* **23**(1): 32–40.

Dobrev, P., Famolari, D., Kurzke, C. & Miller, B. A. (2002). Device and Service Discovery in Home Networks with OSGi, *IEEE Communications Magazine* **40**(8): 86–92.

Golder, S. A. & Huberman, B. A. (2006). Usage patterns of collaborative tagging systems, *J. Information Science* **32**(2): 198–208.

Gu, T., Pung, H. K. & Zhang, D. Q. (2004). Toward an OSGi-based infrastructure for context-aware applications, *IEEE Pervasive Computing* **3**(4): 66–74.

Heckmann, D. (2003). Integrating Privacy Aspects into Ubiquitous Computing: A Basic User Interface for Personalization, *Artificial Intelligence in Mobile System Workshop at Ubicomp (AIMS)*.

Hotho, A., Jäschke, R., Schmitz, C. & Stumme, G. (2006a). Folkrank: A ranking algorithm for folksonomies, *Proc. FGIR 2006*.

Hotho, A., Jaschke, R., Schmitz, C. & Stumme, G. (2006b). Information retrieval in folksonomies: Search and ranking, *The Semantic Web: Research and Applications*, Springer, pp. 411–426.

Hotho, A., Jäschke, R., Schmitz, C. & Stumme, G. (2006). Information retrieval in folksonomies: Search and ranking, *Proceedings of the 3rd European Semantic Web Conference*, LNCS, Springer, Budva, Montenegro, pp. 411–426.

Howes, T. (1996). A String Representation of LDAP Search Filters, *Technical Report RFC 1960*, University of Michigan.

Lee, C., Nordstedt, D. & Helal, S. (2003). Enabling Smart Spaces with OSGi, *IEEE Pervasive Computing* **2**(3): 89–94.

Meyer, S. & Rakotonirainy, A. (2003). A survey of research on context-aware homes, *CRPITS '03: Proceedings of the Australasian information security workshop conference on ACSW frontiers 2003*, Australian Computer Society, Inc., Darlinghurst, Australia, Australia, pp. 159–168.

Michlmayr, E., Cayzer, S. & Shabajee, P. (2007). Add-A-Tag: Learning Adaptive User Profiles from Bookmark Collections, *Proceedings of the 1st International Conference on Weblogs and Social Media (ICWSM'06)*.

Mizzaro, S., Nazzi, E. & Vassena, L. (2009). Collaborative annotation for context-aware retrieval, *ESAIR '09: Proceedings of the WSDM '09 Workshop on Exploiting Semantic Annotations in Information Retrieval*, ACM, pp. 42–45.

Niwa, S., Doi, T. & Honiden, S. (2006). Web page recommender system based on folksonomy mining for itng ’06 submissions, pp. 388–393.

Oh, Y. S., Yoon, H. S. & Woo, W. T. (2006). Simulating Contest-Aware Systems based on Personal Devices, *Proceedings of the International Symposium on Ubiquitous VR*, pp. 49–52.

OSGi Service Platform, Core Specification, Release 4 (2005). The OSGi Alliance.
URL: *http://www.osgi.org*

OWL-S: Semantic Markup for Web Services. 1.1 Release (2004). The OWL Services Coalition.
URL: *http://www.daml.org/services/owl-s/1.1/*

Specia, L. & Motta, E. (2007). Integrating folksonomies with the semantic web, *ESWC '07: Proceedings of the 4th European conference on The Semantic Web*, Springer-Verlag, Berlin, Heidelberg, pp. 624–639.

Strang, T. & Linnhoff-Popien, C. (2004). A context modeling survey, *Workshop on Advanced Context Modelling, Reasoning and Management as part of UbiComp 2004 - The Sixth International Conference on Ubiquitous Computing*.

Wal, T. V. (2007). Folksonomy coinage and definition.
URL: *http://vanderwal.net/folksonomy.html*

Zimmermann, A., Specht, M. & Lorenz, A. (2005). Personalization and context management, *User Modeling and User-Adapted Interaction* **15**(3-4): 275–302.

6

Aging in Place: Self-Care in Smart Home Environments

[1]Blanson Henkemans, Olivier A., [1]Alpay, Laurence, L.,
[1,2]Dumay Adrie, C.M.
[1]*TNO, Quality of Life*
[2]*Delft University of Technology, Institute for Public Sector Efficiency Studies*
The Netherlands

1. Introduction

Older adults prefer living independently at home, in their familiar environment, over a longer period of time (Vermeulen, 2006). Facilitating aging in place helps increasing older adults' quality of life. It enables actively participating in the community and maintaining social networks, increases personal security, and limits the negative effect of relocation, such as confusion, alienation and developing speeding up mental and physical decline. Illustratively, aging in place can contribute to preservation of cognition, decrease in depression, and prevention of incontinence (e.g., Bassuk, et al., 1999; Berg-Warman, 2006; Marek et al., 2005). Maybe most importantly, aging in place offers the possibility to live in an environment which helps older adults to relive old memories dear to them.

In addition, aging in place can contribute to levelling the imbalance in care supply and demand due to the population's aging. The global population is ageing, i.e., the median age rises, and the number of older adults who are requiring more health care increases[1]. Illustratively, in the Netherlands, it is estimated that in 2030, over 4 million people, almost a fourth of the population, will be over 65 years old (Blokstra, 2007). As a result, the number of people with chronic conditions is unceasingly rising. Due to aging, people become more prone to chronic diseases, such as diabetes, heart failure and dementia, often suffer from multimorbidity (i.e., multiple, correlated diseases), over an extended period of time. Also, due to aging, there will be a decrease in the number of potential care professionals, which adds to their increasing workload (Bakker et al., 2005). Furthermore, the shortage in the medical workforce will lead to higher care prices and an increase in the economic burden for patients, insurance companies, government and, thus, for society (Druss et al., 2001). The possible impact of this demographic development raises serious concerns and there is urgency in reducing the gap between demand and supply in health care.

An important requirement for aging in place is the ability for people to perform self-care. Self-care is undertaken by individuals, together with families, and communities with the intention of preventing disease, limiting illness, and restoring health (Leventhal et al., 2004).

---

[1] http://hdr.undp.org/en/media/hdr05_complete.pdf, accessed on October 28, 2009

Examples of self-care activities are maintaining a healthy diet, performing physical activities, using medical instruments, such as glucose and blood pressure meters, setting personal goals and increasing health literacy. Other terms used for increase and maintenance of health are disease self-regulation and self-management, but for this chapter we have chosen the term self-care for its clarity and conciseness.
As illustrated in the following scenario, an older adults' current living environment does not always facilitate aging in place. *Mrs. Brown is an older woman (76) and lives alone and independently at home. Her husband has passed away 5 years before. Mrs. Brown suffers from diabetes type II and recently has developed a light dementia. She prefers to continue living at home to maintain an active and socially participative life. One night, Mrs. Brown did not sleep well. She woke up a couple of times to go to the bathroom and the last time she gets shaky and confused on her way back to bed, possibly due to a hyperglycaemic attack. She panics, trips and falls to the floor. She is not able to send an alarm, because she cannot find the emergency button, nor to find her way back to bed. Her daughter finds her only next morning in a hurt and highly distressed state.* As illustrated in this scenario, aging in place poses a trade-off. On the one hand, aging in place contributes to quality of life aspects, i.e., social, mental and physical wellbeing, and, on the other hand, it increases the burden of the older adults to perform self-care and the need for constant monitoring and response in case of acute need for help.
Smart Homes provide a possibility to address this trade-off and support aging in place (Blanson Henkemans et al., 2007). As defined by Demiris and Hensel (2008), "a 'smart home' is a residence wired with technology features that monitor the well-being and activities of their residents to improve overall quality of life, increase independence and prevent emergencies." They offer a promising and cost-effective way of improving home care for older adults in a non-obtrusive way, which allows greater independence, maintaining good health and preventing social isolation. Smart homes are equipped with sensors, actuators, and/or biomedical monitors. These devices operate in a network connected to a remote centre for data collection and processing (Chan et al., 2009). Accordingly, Smart Homes are becoming increasingly popular and are receiving more and more focus as support environments for healthy, socially participating and self-caring inhabitants (Ackerman, 2009).
When taking into consideration the possibilities of Smart Home environments for aging in place, our research question reads as follows: How can Smart Home environments, with its different support functionalities, contribute to aging in place, with the focus on self-care and support of acute health problems? To answer this question, we elaborate on the importance of self-care for aging in place and its implications for support requirements. Subsequently, we will make an inventory of Smart Home functionalities and describe their main characteristics in relation to supporting aging in place. Finally, we will discuss how Smart Homes facilitate aging in place. This chapter concludes with a discussion on how self-care activities and support acute help are supported by applying existing Smart Home functionalities and what the remaining challenges are.

2. Aging in Place: Self-care for Independent Living

As discussed in the Introduction section, an important aspect in aging in place is the possibility for older adults to perform self-care. People with chronic diseases, such as diabetes, chronic hart failure, and dementia, are faced with psychosocial problems, such as stigmatism; moreover, they need to manage daily living according to their financial and social conditions to be able to live independently at home. Furthermore, patients require sufficient knowledge about their condition and its treatment, performance of condition management activities, and application of the necessary skills to maintain adequate psychosocial functioning (Barlow et al., 2002). Accordingly, self-care aims at educating and increasing the patient's intrinsic motivation, which in turn can lead to stimulation of maintaining a healthy lifestyle and adhering to medical treatment (Maes & Karoly, 2005). For this, patients need to be provided with relevant medical information and support for developing self-care skills, i.e., problem solving, decision making, resource utilization, forming of a patient-caregiver partnership, and taking action (Lorig & Holman, 2003).

A prerequisite for self-care is the empowerment of patients. Patients should be able to perform self-care activities themselves, but also need to be facilitated by their informal and formal caregivers, such as peer and home care, specialist, and nurse practitioner. Caregivers should empower patients to take their care and health more in their own hand. Whether self-care means taking medication, making life-style changes or undertaking preventive actions, patients must be guided to make their own day-to-day decisions (Newman et al., 2004). This implies allocating responsibility to the patient (Alsop & Heinsohn, 2005), providing education about the illness and treatment options, finding out practical information on healthcare services, and understanding the medical jargon (Barlow et al., 2002). Accordingly, patients and caregivers operate more on the same level, which increase the capabilities of the patient and independence long term.

Based on these theories of self-care and patient empowerment, we define four aspects of self-care, which are the following (Blanson Henkemans et al., 2010):

1. Gaining a good insight in their own health condition;
2. Retrieving personal information to support choices of self-care activities;
3. Fitting self-care activities in daily life and developing self-care habits;
4. Involving the environment to support self-care.

In addition, it is essential for people to maintain a healthy lifestyle, but equally to maintain a good quality of life, i.e., enjoy their social and professional life, have room for personal interests (e.g., hobbies), and maintain a good psychological well being. We will now elaborate further on these four aspects.

First, patients need to gain insight in their own condition. For example, patients with diabetes, such as Mrs Brown in our scenario, are confronted with a great variety of data about their condition. Nutrition, physical activity, monitoring results, and medication are all details related to someone's health condition. Moreover, it is important for the patient to keep track of the medical condition, current health and the prognosis (Halme et al., 2005). More specifically, keeping track of their diet, physical activities, medication prescription, intake and actual working, and arranging the entries and sharing them with relevant actors, such as family and caregivers, help people to better understand their current conditions and possibly to foresee possible upcoming complications.

Second, people need to acquire information about their condition, such as its background, consequences and treatment possibilities, and develop health literacy (Murray et al., 2004b). Based on this information and insights in their own condition, patients can evaluate and choose between different personally preferred self-care options (e.g., change of life style, medication intake, surgical operation, or non-treatment). The choice is based on the consequence of their choice on important life domains, such as work, family, social life and hobbies. Information can help structure the possibilities and help making the most fitted choice (Thun et al., 2008; Woloshin et al., 2000). Specifically, older adults who are in receipt of information geared to their personal characteristics (e.g., medical history, phase of disease, treatment, and norms and values), will translate their condition insights into self-care activities better. They can weigh off the consequences of certain self-care choices on their health against their constraints for important life domains, such as participating in (voluntarily) work and social activities and hobbies.

Third, the selected self-care activities need to be developed into habits, combined with daily activities. Here, translating a treatment to personal goals and setting plans to achieve those goals are essential (Lorig et al., 2006). People want to adhere to their treatment, e.g., treat their diabetes, which requires the motivation to perform certain activities, such as exercising on a daily basis. This implies that they have to internalize their treatment (Deci & Ryan, 2002), i.e., determine their preferences and intentions, problem solve probable pitfalls, and then make their activities specific, measurable, acceptable, realistic and time bound (SMART) (O'Neill et al., 2006). Illustratively, Mrs Brown, who decides to perform exercise regularly can plan an half hour mall walk, on every Monday, Wednesday and Saturday, which are the days mall walks occur and her daughter or neighbour is available to pick her up because she goes shopping on those days. After three months she checks if she indeed participated, on average, in two walks a week and what effect it had on her HbA1c (i.e., average glucose level over a period of three months) and general physical mobility. The latter implies monitoring and providing feedback which, in line with the motivational interviewing approach, applies an empathetic perspective and focuses on increasing self-efficacy, stimulating self-reflection and offering guidelines for independently overcoming challenges, which can contribute to realizing these self-care habits (Rollink et al., 2008).

In the discussion on self-care activities, we already mentioned repeatedly the importance of the environment. Involving family members and neighbours is good example, as is joining clubs in the community that can give social and possibly, health support. Also peer, home and specialist care need to be involved. To realize this, the older adults need to organize the support of partners, family members, neighbours, community centres and informal and formal care. For this organization, acquiring information about possible support is essential. Moreover, it will contribute most to aging in place when the support is provided integrally, i.e., each actor plays a role, collaboratively, in the care of the patient and the support of self-care (Molema, 2009).

In relation to patient empowerment, the older adult in a home situation is better positioned to be aging in place, when three main requirements are met, which are:

- Access to information about illnesses and diseases, available health care services and care options and to tools such ICT-based applications or medical devices;
- Skills to carry out self-care activities (such as problem solving, understanding, decision making);
- Motivation to achieve one's own goals. Information and services tailored to the person's own health situation (personalized and contextualized) are also intertwined to these requirements.

These requirements of empowerment are facilitated via a number of patient-centred mechanisms, which enable communication, provide education and raise the level of health literacy, provide information, enable and support self-care practices, provide decision-aids, support contact with fellow patients, and facilitating integrated care (Monteagudo & Moreno 2007; Aujoulat et al., 2008).

Characteristic	Definition	Example	Possible implications for aging in place support
Sensation	The awareness of simple properties of stimuli, such as colour; activation of sensation cells, e.g., retinal cells)	Recognizing the colour red, hear high frequencies	Overlooking important notifications (e.g., dangerous fluid, doorbell) due to obscured imagery of or inaudible noise
Perception	The awareness of complex characteristics of things in the environment; the interpretation of information that results forms sensation.	Identifying a red object as an apple or recognizing an sounds as an alarm	Misunderstanding a complex combination of sound, light and tactile signals and performing incorrect activities accordingly. For example, misinterpreting an intense sound and visual alarm for a high glucose measurement for a low one.
Cognition	Process by which the brain takes sensory information from ears, eyes, etc. and transforms, reduces, stores, recovers and uses that information	Thinking, problems solving, reasoning en making decisions	Difficulties to grasp a multitude of health data and infer necessary self-care activities; recalling required steps to accurately use tool to self-monitor biomarker (e.g., glucose level)
Movement control	Carrying out an action based on perception or cognition; requires coordination of muscles for control of motion of some type.	Driving a car, double click a Mouse button, picking a book of the shelf	Lack of ability to manipulate an interface, due to decline of dexterity, which makes (swift) use of technology more challenging.

Table 1. Description and examples of age related personal characteristics, as posed by Fisk et al. (2009), and their possible implications for aging in place support.

Finally, older adults are faced with dynamic age related factors, such as cognitive, sensory, perception, and movement skills, and with the side effect of their health condition. This poses additional challenges in living independently, with the use of technical support offered (Fisk et al., 2009), and as a result may constrain the possibilities of aging in place in Smart Homes. Table 1 presents a description and examples of the different age related factors and gives possible constraints for aging in place support. As an illustration of the effect of age and condition related characteristic on aging in place, older adults with diabetes type II can experience difficulties remembering their medication, which possibly leads to an acute hyperglycaemic attack. When offering technical support in the form of visual and additive support (e.g., computer display providing textual instructions about what to do, accompanied with speech), both dizziness, shakiness and confusion, as an effect of the attack, and cognitive overload, due to multitude of multi-perceptive modalities, can negatively affect the usability of the support. Consequently, when supporting self-care activities and offering acute help, it is important to take into consideration the personal requirements of the older adults to realize success with the offered supporting mechanisms, such as alarms, complex notifications with large amounts and layered information and interfaces that require detailed manipulations.

In summary, to support aging in place, it is important to facilitate self-care and promote patient empowerment. Patients need to be able to gain insight in their condition, acquire information related to their condition and their personal preferences, translate their treatment to personal life style goals, and involve the environment in their self-care. Patient empowerment, which focuses on making accessible of personal information about health and care, training skills to cope with the illness and motivating to set and achieve own goals, plays an important role in facilitating self-care. Finally, age related factors, i.e., cognitive, sensory, perceptive and motor skills, impact how well support is geared to personal requirements and need to be taken into consideration when offering support for aging in place, relating to communication, education information, decision-aids, fellow patients forums, and integrated care.

3. Smart Home Initiatives

Demiris & Hensel (2008) gave an elaborate systematic review of health related Smart Home projects. Their international inventory (i.e., United States, Europe and Asia) covered 21 smart home initiatives, including the Aware Home Research Initiative at the Georgia Institute of Technology; Place Lab, Massachusetts Institute of Technology, ENABLE project, which is a joined initiative from UK, Ireland, Finland, Lithuania and Norway, Philips Care Lab, Eindhoven, the Netherlands; PROSAFE in Toulouse, France; Welfare Techno-House project, Japan. Across these different initiatives, the authors identified the following functionalities present in Smart Homes:

1. Physiological monitoring of physiological measurements (e.g., pulse, respiration, temperature, and blood pressure, as well as blood sugar level);
2. Functional monitoring of functional measurements (e.g., motion, meal intake, and other activities-of-daily-living, whereby abnormal or critical situations (e.g., falls) are detected;
3. Safety monitoring and assistance of environmental hazards (e.g., fire or gas leak). Assistance includes automatic turning on and off bathroom lights when getting out of bed and facilitating safety by reducing trips and falls;
4. Security monitoring and assistance of domestic threats (e.g., intruders). Assistance includes notification of external relevant actors;
5. Social interaction monitoring and assistance of social interactions (e.g., phone calls, visitors, and participation in activities). Assistance includes technologies that facilitate social interaction (e.g., video-based components that support video mediated communication with friends and loved ones and virtual participation in group activities);
6. Cognitive and sensory assistance of automated or self-initiated reminders and other cognitive aids for users with identified memory deficits (e.g., medication reminder and management tools, lost key locators). Aids include task instruction

technologies (e.g., verbal instructions in using an appliance) and aids for sensory deficits (e.g., sight, hearing, and touch).

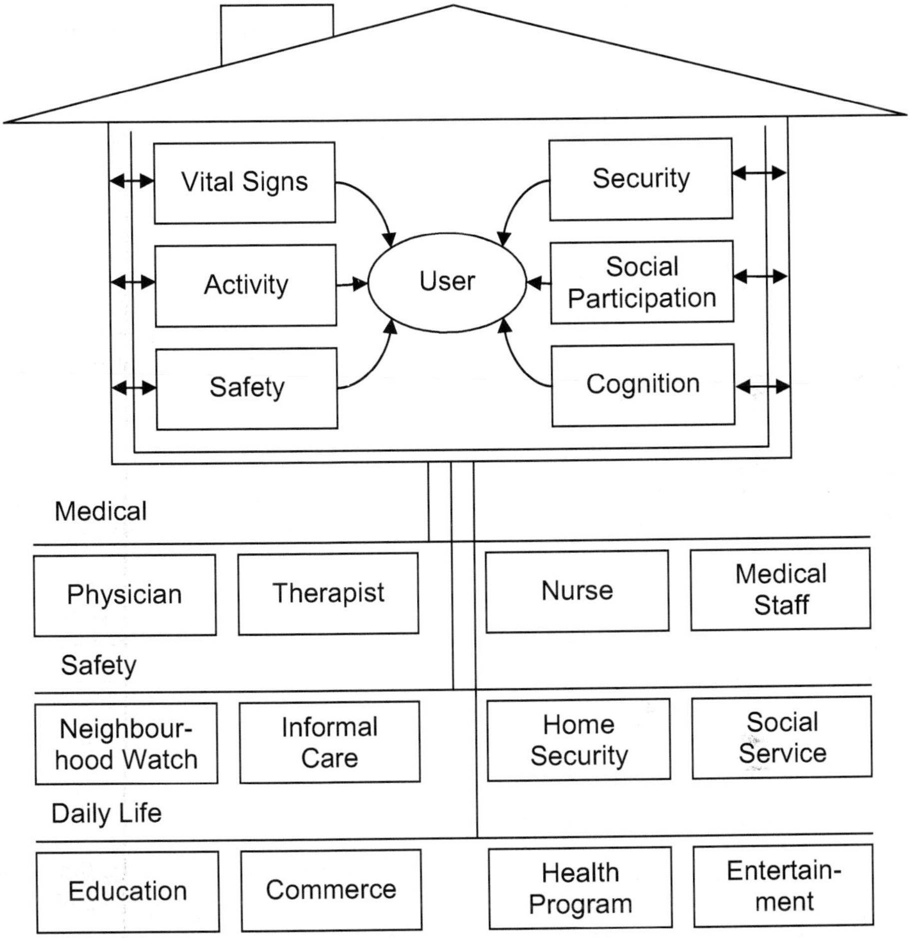

Fig. 1. Key organization in Smart Home, based on Chan et al. (2009) and Demiris & Hensel (2009).

By offering these functionalities, integrated as illustrated in Figure 1, Smart Homes pose interesting benefits. Chan and colleagues (2009) point out the following possibilities of Smart Homes. First, through the different monitoring possibilities, they enable measurements of vital signs and behavioural patterns, which can be translated into accurate predictors of health risks and combined with alarm-triggering systems to initiate appropriate medical action. Second, the monitored data can support transmural care (Celler et al., 2003). That is to say, they facilitate an infrastructure for coordinating multidisciplinary care outside the hospital (scheduling visits with health staff and community health workers, automating collection of clinical findings and test results) and providing means for nursing services in the home (Finch et al., 2008). Third, through cognitive, sensory, and socially participation monitoring and assistance, Smart Homes can stimulate patient-centred care (Paré et al., 2007). The offered functionalities enable a patient-centred management approach that provides accurate and reliable data, empowers patients, influences their attitudes and behaviours, and potentially improves their medical condition. By providing accurate and up-to-date information, to help take better decision, patients become more responsible, informed, expert, and educated self-managers. Fourth, Smart Homes offer economic benefits. The use of functionalities in the form of, for example, personal health records (PHR), e-prescribing, decision support systems, electronic management of chronic illness, can contribute to increasing care efficiency. This is due to time and cost reduction, reducing care need through prevention of illness deterioration, and supervising and early establishing medical errors (Anderson, 2007). Finally, Smart Home Labs, such as the Georgia Tech Aware Home, offer a testing ground for generic environmental constructs and their measurement, as well as a unique setting from which new understandings of person-environment fit can emerge. Essentially, Smart Homes offer a domestic environment for natural use of technology in a controlled and observational environment (Blanson Henkemans et al., 2007).

Another interesting development in Smart Homes is the use of robotics. In the domain of health care, robotics can be deployed for five objectives (Butter et al., 2008):

- Assisting in preventive therapies and diagnostics, through robotized analysis of motion and coordination, intelligent fitness systems, tele-diagnostic and monitoring robotics systems, and smart medical capsules;
- Facilitating people with disabilities and chronic conditions to continuously perform their daily activities, through robotized systems supporting manipulation and mobility, and intelligent prosthetics;
- Assisting professional care giving, through robotized logistical aids for nurses, patient monitoring systems, physical tasks in care provision, and paramedic tasks;
- Rehabilitating patients, through robot-assisted motor-coordination, physical training and mental cognitive and social therapies;
- Supporting surgery, through robot-assisted precision, minimal invasive and remote surgery, and medical micro- and nanobots.

Due to this wide range of application possibilities, robotics can contribute to reducing labour costs, increasing independence social independent, quality of care and, the performance of activities otherwise not executable for humans, such as lifting heavy weights.

In addition, due to their physical appearance, they are able to offer assistance in a social intelligent manner (Looije et al., 2009). A social intelligent robot that offers three support roles, i.e., educator, motivator and buddy, can respectively inform patients with diabetes about their illness, guide self-care goal-setting, and offer empathetic feedback to help attaining them. Moreover, by assessing personal characteristics, recalling previous interactions, and having a social dialogue (with gaze, facial expression, and vocal intonations), the robot could develop an inspiring relationship with the patients (Blanson Henkemans et al., 2009).

An aspect we need to address, when stating the advantages of Smart Homes is the deficiency in empirical research. When we look at the reviews of Demiris and Hensel (2008) and Martin et al. (2008), a number of shortcomings are found in current research. Most of the studies are pilot or short-term projects, consisting of nonrandomized trials without control groups, which often show methodological weaknesses (e.g., in samples size, context, and study design), limiting the generalization of the findings. Also, the few randomized controlled trials that are conducted refrain from comparing Smart Home interventions with conventional health care practices. Accordingly, although the current literature depicts great potential of Smart Homes for aging in place, it is currently difficult to accurately point out their clinical and economical benefits.

4. Smart Homes for Aging in Place: Application and Challenges

In the future, Smart Homes can add to performing self-care and, accordingly, to aging in place. An integrated system of different functionalities, which monitor and assist psychological and functional functioning, safety, security, social interactions and providing cognitive and sensory assistance, will be offered through various devices, such as Information and Communication Technology (ICT) and robotics (e.g., assistive device, robotic-assistant, companion robot, autonomous wheelchair, and stair lift). These devices are capable of providing support for making decisions and diagnoses, improve inhabitants' access to health care services and optimizing resource utilization, control of home appliances, such as heating, air-conditioning, windows, and stoves.

In addition, when we look at transmural care support for aging in place, Smart Homes are connected to hospital, which increasingly become a central health information centre. Accordingly, these centres direct activities in the Smart Homes, supported by technology on location, which decreases the requirement for people to be hospitalized for their condition. For example, hospital-based health professionals can initiate consults online, i.e., eConsult, and make virtual visits to the patient, and specialized care tasks can be reallocated to mobile health professionals with technical support on location.

4.1 Application of Smart Homes for Aging in Place

As illustrated in Figure 2, following are a number of Smart Home applications for aging in place. People with dementia can be navigated by a pet robot to find their bathroom during the night and return safely to bed. Moreover, the bathroom contains a number of sensors that can monitor vital functionalities, which enables diagnosis and possible early detection of physical complications. The collected data is stored in a PHR, managed by the inhabitants and (remotely) accessible for the people they feel appropriate. Movement sensors are set throughout the house, which detect movement and can infer unusual patterns, or lack of movement. For example, when a person falls and stays immobile on the floor, a notification can be sent to neighbours, family or relevant caregivers.

eConsults are made possible through personal computers and on large screens throughout the house with interface specifications geared to aging related characteristics (e.g., big font, recognizable colours, singular perceptive modalities, and information provided in small doses). People in Smart Homes can virtually consult with caregivers, which includes elaborate educational modules that provided information related to the treatment and topics discussed. Because the service is accessible when relevant and convenient for the patient, it can increase personalized access to information and training of skills to translate prescribed treatment to self-care activities. Virtual meetings can also be put into practice to communicate with family, people in the community, peers, and even a computer coach, i.e., an eCoach, for empathetic motivating and educational support and also entertainment, which can increase the quality of life. The eCoach can be presented on the computer and as mobile social intelligent robot.

In addition to their social role, robots can also help with daily household activities, which become challenging for the inhabitants, enabling them to stay longer independently in their home. Exemplary activities are cleaning, filling and emptying the dish washer, supporting mobility in the house (e.g., walking the stairs) and outside the house (e.g., gardening, doing groceries, and meeting people in the neighbourhood). Moreover, they can support home care with their activities, such as lifting people out of bed.

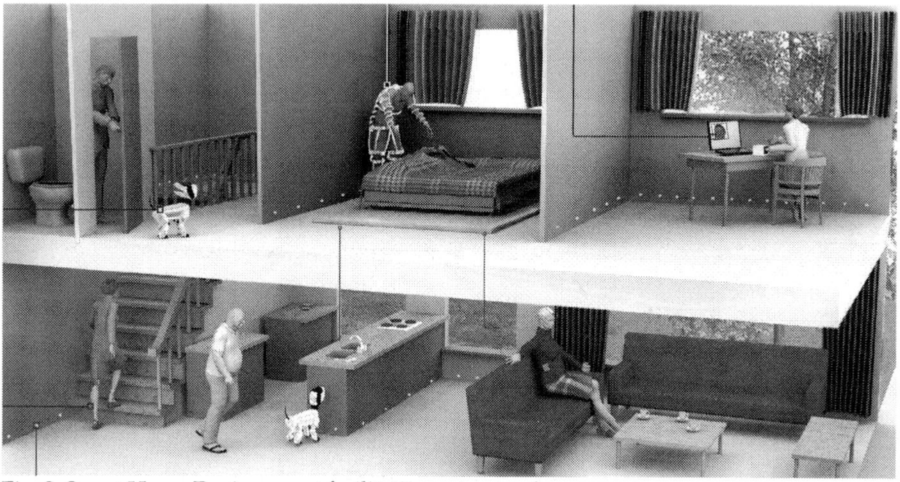

Fig. 2. Smart Home Environment facilitating aging in place.

4.2 Challenges for Smart Homes for Aging in Place

Despite the vast proposed benefits Smart Homes have to offer, there are still a number of challenges in relation to supporting self-care activities and, thus, aging in place. The first challenge is the standardization of the technology in the environment. Professionals outside a networking system are faced with lack of ability to exchange clinical data with laboratories and hospitals. Also, because the industry that delivers the functionalities for Smart Homes tends to be dominated by suppliers the common approach is technology-push, rather than a demand-pull approach, which causes lack of user centred design (UCD) and, thus of user friendly applications.

As we saw in Table 1, gearing to users' needs is specifically important when supporting older adults with aging in place, since they are faced with dynamic sensory, perceptive, cognitive and movement skills, and standard support technology may greatly neglect their personal requirements (Fisk et al., 2009). Lack of usability leads to decline in self-efficacy and to mistrust in relation to the technology, which in turn elicits breached privacy feelings.

Introducing technology in the house could trigger several issues: accidental disclosure of individuals' data, contacting the wrong people, and incorrect use of data (Croll & Croll, 2007). As a result, in case of mistrust, the inhabitants of Smart Homes may decide to withhold information, disclose obscured data to health care providers, or avoid using the health care support system altogether, which goes at the cost of the effectiveness of the Smart Homes' functionalities. Lack of usability may also lead to the conception that technology will replace personal interaction with their health care providers and they may worry about a technology affecting their lifestyle, financial status, emotional and psychological wellbeing of family members (Bauer, 2001).

In summary, UCD needs to be a constant factor in and throughout the development of technology and implementation in Smart Homes (Vredenburg et al., 2002). Users, including inhabitants, people in their environment, caregivers and stakeholders, such as hospitals, insurance companies, industry, and policy makers, are involved early in setting up of design specification, designing and evaluating prototypes, and in the implementation process (Blanson Henkemans et al., Submitted). The latter implies that the goal and background of the technology is explained in a way understandable for the user, whereby special attention goes out to ethical issues (Bauer, 2001), and its introduction is elaborately guided.

5. Discussion

People prefer living longer and independently at home, but aging in place poses a thorny trade-off. On the one hand, it contributes to older adults maintaining a mental, physical and social wellbeing and adding to their quality of life. On the other hand, because older adults are often faced with one or multiple chronic conditions, their wish to age in place compels them to perform complex self-care activities, preventing disease, limiting illness, and restoring health. Moreover, there is a need for constant monitoring in case of acute need for help, i.e., when health threatening situations occur. Smart homes, i.e., residences containing technology that monitor the well-being and activities of their residents, become increasingly popular and receive more focus as support environments for healthy, socially participating and self-caring inhabitants (Demiris & Hensel, 2008; Ackerman, 2009). Accordingly, in this chapter we studied how Smart Home environments, with its different support functionalities, can contribute to aging in place, with the focus on self-care and support of acute problems with their wellbeing.

In line with the research on self-care (e.g., Barlow et al., 2002; Leventhal et al., 2004; Lorig et al., 2003; Maes & Karoly, 2005), we elaborated on the importance of self-care for aging in place and its implications for support requirements. Based on these theories, we defined four main self-care activities for older adults aging at home, which are: gaining a good insight in the personal health condition, retrieving personal information to support choices in self-care activities; fitting self-care activities into daily life and developing healthy habits; involving the environment to support self-care. To increase the chances for people to actually continuously perform these self-care activities, it is essential to facilitate combining a healthy lifestyle with a good quality of life, i.e., enjoy their social and professional life, have room for personal interests (e.g., hobbies), and maintain a good psychological well being (Blanson Henkemans et al., 2010). Moreover, both health condition, e.g., shakiness and aggravation, and age related factors, i.e., cognitive, sensory, perceptive and motor skills, impact how well Smart Home functionalities are geared to personal requirements and need to be taken into consideration when offering support for aging in place.

When looking at the literature (e.g., Demiris & Hensel, 2008; Martin et al., 2007; Chan et al., 2009), Smart Homes offer the possibility to monitor and assist physiological and functional activities, safety, security, and social activities. Also, they offer cognitive and sensory assistance. These functionalities are offered by integrated technology, such as computers, databases, sensors, video cameras, interfaces (e.g., monitors) implemented ubiquitously in and around the house and in connection with remote actors, such as family, caregivers and other supervising units. In addition, robots active in the Smart Home environment offer the possibility of mobile guidance and physical support (Butter et al., 2009).

An important benefit of Smart Homes for aging in place is online monitoring of the inhabitants' vital, movement and general wellbeing data and their transfer to supporting actors, such as family, peer and home care, and specialists. Also, with technology on location, caregivers become less reliant on the hospital or clinic, making them more mobile, enabling multidisciplinary care in the home (Finch et al., 2003), and facilitating patient-centred care. Illustratively, in Smart Homes, care can be offered virtually (e.g., eConsult), remotely coordinated (e.g., nursing service remotely guided by specialists) and directly (e.g., professional linked to the hospital, which functions as a central health information centre). The appropriate type of care can be selected depending on the health condition and preferences of the patient. Finally, with the use of robotics, daily activities that become challenging for older adults can be supported by robots. For example, emptying the dish washer can be fully allocated to the robot assistant, doing gardening can be supported by an exoskeleton robot that facilitates continuous mobility, and going to the bathroom at night can be guided by a robotic pet dog, with directing spotlights and distinct noises.

Challenges that need to be overcome lay in the realm of experienced usability of the Smart Home technology by the inhabitants, their environment and the caregivers. When they are insufficiently involved in the development and implementation process, caused by a technology-push approach (Barlow et al., 2006), users may mistrust the technology in their house and question the ethics, considering the strong focus on monitoring. This may lead to obscured data and to ineffective Smart Home support, accordingly. Another result is that people feel that technology is posed upon them and may replace personal interaction with their social environment and health care providers. The technology is foremost a facilitating tool to enable aging in place by complementing human care and not by replacing it. Only when it is apparent to the users, including the caregivers, that the technology is there to meet with this facilitating function and fit with their daily life objectives, e.g., participating in social activities, maintaining hobbies, and receiving and providing personal care, will Smart Homes be fully adapted.

Strikingly, although indicated as one of the possible benefits by Paré et al. (2007), the established Smart Home functionalities offer little concrete support to elicit intrinsically motivated self-care activities. The main focus lies in monitoring and offering accurate and up-to-date information to help older adults to make better decisions and become more informed. This may indeed help managing medical and wellbeing data, receiving personalized information and involving actors the environment with self-care activities. However, besides the use of social intelligent robots as motivating partners (Looije et al., 2009), little attention goes out to translating treatment to personal self-care goals, to planning, attaining, and maintaining them, and to iterative provision of empathetic, self-reflective and empowering feedback. As a result, with the currently developed Smart Home functionalities, people may understand their condition and decide what self-care activities to perform short term (e.g., maintaining healthy diet, exercising regular, and taking medication) and may be met in their acute care needs, but will experience challenges to actually put it to practice long term and develop healthy habits required to age in place (Deci & Ryan, 2002).

The theory on self-care, as described in this chapter, provides useful guidelines to develop the functionality for developing healthy habits, but assessing how practically developing and implementing it asks for an extension in user-centred and empirical research in relation to the effect of Smart Homes on quality of life (e.g., functioning, emotional well-being, social involvement, and satisfaction), clinical outcomes, and financial benefits (Vredeburg et al., 2005; Cutler, 2007; Gitlin, 2003). Interestingly, Smart Homes, with their monitoring facilities, offer great testing ground possibilities. Of course, as with monitoring in non-experimental Smart Homes, special attention needs to go out to the (medical) ethical issues, such as the guarantee of voluntary participation and a good balance between participants' risks and social and scientific benefits.

When augmenting current monitoring and assisting functionalities with support of long term self-care, in regards to setting, attaining and maintaining personal self-care goals, Smart Homes offer great potential for aging in place. As illustrated in the following scenario, this can help increasing quality of life, by enabling actively participating in the community and maintaining social networks, increases personal security, and limits the negative effect of relocation (e.g., Berg-Warman, 2006; Marek et al., 2005). Moreover it can contribute to levelling the forecasted imbalance in health care demand and supply, as described in the following scenario. *Mrs. Brown experiences some problems with her glucose level and needs to go to the bathroom multiple times throughout the night. The sensors in her Smart Home detect her movements and instruct the pet robot to physically guide her to and from the bathroom. Based on the detection of unbalanced movements and vital signs (e.g., excessive perspiration), her eCoach suspects a health issue (e.g., glucose level too high) and instructs her service robot to measure her glucose levels and administrate insulin accordingly. The dosage and intervention time are registered in her PHR. Also, both her daughter and the personal diabetes nurse are notified of the occurrence, including the severity, and the latter comes by to check with Mrs. Brown the next morning. She compliments her on how well it goes with her living independently. Later that day, her daughter visits and together they go for a healthy walk to the botanical gardens.*

6. References

Ackerman, M.J. (2009). The Smart Home. *The Journal of Medical Practice Management, 25*, 68-69.

Alsop, R. & Heinsohn, N. (2005). *Measuring empowerment in practice : structuring analysis and framing indicators.* Washington, D.C.: World Bank, Poverty Reduction and Economic Management Network, Poverty Reduction Group.

Anderson J. G. (2007). Social, ethical and legal barriers to E-health. *International Journal of Medical Informatics,* 76, 480–483.

Aujoulat, I., Marcolongo, R., Bonadiman, L., & Deccache, A. (2008). Reconsidering patient empowerment in chronic illness: a critique of models of self-efficacy and bodily control. *Social science & medicine, 66,* 1228-1239.

Bakker DH de; Polder JJ; Sluijs EM; Treurniet HF; Hoeymans N; Hingstman L; Poos MJJC; Gijsen R; Griffioen DJ; Velden LFJ van der. (2007). *Op een lijn - Toekomstverkenning eerstelijnszorg 2020 (Public health forecast for primary care in the Netherlands in 2020).* URL: http://www.rivm.nl/bibliotheek/rapporten/270751009.pdf, accessed on November 3, 2009.Barlow, J., Bayer, S., & Curry, R. (2006). Implementing complex innovations in fluid multi-stakeholder environments: Experiences of 'telecare'. *Technovation, 26,* 396-406.

Barlow, J. H., Sturt, J., & Hearnshaw, H. (2002). Self-management interventions for people with chronic conditions in primary care: Examples from arthritis, asthma and diabetes. *Health Education Journal, 61,* 365-378.

Bassuk, S. S., Glass, T. A., & Berkman, L. F. (1999). Social disengagement and incident cognitive decline in community-dwelling elderly persons. *Annals of internal medicine, 131,* 165-173.

Bauer, K. A. (2001). Home-Based Telemedicine: A Survey of Ethical Issues. *Cambridge quarterly of healthcare ethics : CQ : the international journal for healthcare ethics committees., 10,* 137-146.

Berg-Warman, A. B. (2006). The supportive community: A new concept for enhancing the quality of life of elderly living in the community. *Abstracts in social gerontology., 49,* 69.

Blanson Henkemans, O. A., Boog, P. J. M. v. d., Lindenberg, J., Mast, C. A. P. G. v. d., Neerincx, M. A., & Zwetsloot-Schonk, J. H. M. (2009). An Online Lifestyle Diary with a Persuasive Computer Assistant Providing Feedback on Self-Management. *Technology and Health Care Special Issue "Smart environments: technology to support healthcare", 17,* 253-267.

Blanson Henkemans, O. A., Molema, J. J. W., Alpay, L. L., Schoone, M., Otten, W., Boog, P. J. M. v. d. et al. (2010). Innovaties voor Zelfzorg: Ontwikkelen van Kennis, Diensten en Technologie voor Duurzame Gezondheidszorg (Innovations for Self-Care: Development of Knowledge, Services and Technology for Sustainable Care). *Tijdschrift voor Gezondheidswetenschappen (TSG).*

Blanson Henkemans, O. A., Molema, J. J. W., Alpay, L. L., & Dumay, A. M. C. (Submitted). Sustainable eHealth Services for Mulimorbidity in Complex Health Care Networks. In *13th International Congress on Medical Informatics.*

Blanson Henkemans, O. A., Caine, K. E., Rogers, W. A., Fisk, A. D., Neerincx, M. A., & Ruyter, B. d. (2007). Medical Monitoring for Independent Living: User-Centered Smart Home Technologies for Older Adults. In *Med-e-tel 2007* Luxemburg, Luxemburg.

Blokstra, A. (2007). *Vergrijzing en toekomstige ziektelast : prognose chronische ziektenprevalentie 2005-2025.* Bilthoven: RIVM.

Butter, M., Rensma, A., Boxsel, J. v., Kalisingh, S., Schoone, M., Leis, M. et al. (2008). *Robotics for Healthcare.* European Commission: Information Society and Media.

Celler, B. G., Lovell, N. H., & Basilakis, J. (2003). Using information technology to improve the management of chronic disease. *The Medical journal of Australia, 179,* 242-246.

Chan, M., Campo, E., Esteve, D., & Fourniols, J.-Y. (2009). Smart homes - Current features and future perspectives. *Maturitas Maturitas, 64,* 90-97.

Croll, P. R. & Croll, J. (2007). Investigating Risk Exposure in e-Health Systems. *International journal of medical informatics, 76,* 460-465.

Cutler, L. J. (2007). Physical Environments of Assisted Living: Research Needs and Challenges. *The Gerontologist, 47,* 68-82.

Deci, E. L., & Ryan, R. M. (2002). *Handbook on self-determination research: Theoretical and applied issues.* Rochester, N.Y.: University of Rochester Press.

Demiris, G. & Hensel, B. K. (2008). Technologies for an aging society: a systematic review of "smart home" applications. *Yearbook of medical informatics,* 33-40.

Druss, B. G., Marcus, S. C., Olfson, M., Tanielian, T., Elinson, L., & Pincus, H. A. (2001). Comparing The National Economic Burden Of Five Chronic Conditions. *Health Affairs -Millwood Va Then Bethesda Ma-.* 20, 233-241.

Finch, T. L., Mort, M., Mair, F. S., & May, C. R. (2008). Future patients? Telehealth care, roles and responsibilities. *Health and Social Care in the Community, 16,* 86-95.

Fisk, A. D., Rogers, W. A., Charness, N., Czaja, S. J., & Sharit, J. (2009). *Designing for Older Adults: Principles and Creative Human Factors Approaches.* (2 ed.) Boca Raton, FL: CRC Press.

Gitlin, L. N. (2003). Conducting Research on Home Environments: Lessons Learned and New Directions. *The Gerontologist, 43,* 628-637.

Halme, M., Hrauda, G., Jasch, C., Kortman, J., Jonuschat, H., Scharp, M. et al. (2005). *Sustainable Consumer Services: Business Solutions for Household Markets.* London; Sterling, VA: Earthscan.

Leventhal, H., Halm, E., Horowitz, C., Leventhal, E., & Ozakinci, G. (2004). Living with Chronic Illness: A Contextualized, Self-Regulation Approach. In S.Sutton, A. Baum, & M. Johnston (Eds.), *The Sage Handbook of Health Psychology.* London: Sage.

Looije, R., Neerincx, M. A., & Cnossen, F. (2009). Persuasive robotic assistant for health self-management of older adults: Design and evaluation of social behaviors. *International Journal of Human-Computer Studies.*

Lorig, K. R. & Holman, H. (2003) Self-management education history: Definitions, outcomes and mechanisms. *Annals of Behavioral Medicine, 26,* 1-7.

Lorig, K. R., Ritter, P. L., Laurent, D. D., & Plant, K. (2006). Internet-based chronic disease self-management: a randomized trial. *Medical care, 44,* 964-971.

Maes, S. & Karoly, P. (2005). Self-Regulation Assessment and Intervention in Physical Health and Illness: A Review. *Applied Psychology An International Review, 54,* 267-299.

Marek, K. D., Popejoy, L., Petroski, G., Mehr, D., Rantz, M., & Lin, W. C. (2005). Clinical outcomes of aging in place. *Nursing research, 54,* 202-211.

Martin, S., Kelly, G., Kernohan, W. G., McCreight, B., & Nugent, C. (2008). Smart home technologies for health and social care support. *Cochrane database of systematic reviews (Online).*

Molema, J. J. W. (2009). *Hospital System Design: Creating Supply Flexibility to Match Demand Variability.* PhD University of Maastricht, Maastricht.

Monteagudo, J. L., & Moreno Gil, O. (2007). *E-Health for patient empowerment in Europe = E-Salud para la potenciación de los pacientes en Europa.* Madrid: Instituto de Salud Carlos III, Área de Telemedicina y eSalud. URL: http://ec.europa.eu/information_society/activities/health/docs/publications/eh_era-patient-empower.pdf, accessed November 3, 2009.

Murray, E., Burns, J., See, T. S., Lai, R., & Nazareth, I. (2004a). Interactive Health Communication Applications for people with chronic disease. *Cochrane database of systematic reviews (Online).*

Murray, M. D., Wu, J., Tu, W., Clark, D. O., Weiner, M., Morrow, D. G. et al. (2004b). Health literacy predicts medication adherence. *Clinical Pharmacology and Therapeutics, 75,* 76.

Newman S, Steed L, & Mulligan K. (2004). Self-management interventions for chronic illness. *Lancet.* 364 (9444), 23-29.

O'Neill, J., Conzemius, A., Commodore, C., & Pulsfus, C. (2006). *The Power of SMART goals: Using goals to improve student learning.* Bloomington, IN: Solution Tree.

Pare, G., Jaana, M., & Sicotte, C. (2007). Systematic Review of Home Telemonitoring for Chronic Diseases: The Evidence Base. *Journal of the American Medical Informatics Association : JAMIA., 14,* 269.

Rollnick, S., Miller, W. R., & Butler, C. C. (2008). *Motivational interviewing in health care helping patients change behavior.* New York, New York: Guilford.

Thun, M. J., Hannan, L. M., & Stefanek, M. (2008). Risky Business: Tools to Improve Risk Communication in a Doctors Office. *Journal of National Cancer Institute, 100,* 830-831.

Vermeulen, J. N. A. M. (2006). *Langer zelfstandig wonen en hoe ICT daarbij kan helpen : actieonderzoek naar de inzet van ICT ter ondersteuning van ouderen die zo lang mogelijk zelfstandig willen wonen.* Drukkerij Universiteit van Tilburg, Tilburg.

Vredenburg, K., Isensee, S., & Righi, C. (2002). *User-centered design : an integrated approach.* Upper Saddle River, NJ: Prentice Hall PTR.

Woloshin, S., Schwartz, L. M., Byram, S., Fischhoff, B., & Welch, H. G. (2000). A new scale for assessing perceptions of chance: a validation study. *Medical decision making : an international journal of the Society for Medical Decision Making, 20,* 298-307.

7

Telemonitoring of the elderly at home: Real-time pervasive follow-up of daily routine, automatic detection of outliers and drifts

Yannick Fouquet, Céline Franco, Jacques Demongeot, Christophe Villemazet and Nicolas Vuillerme
TIMC-IMAG Laboratory, UMR UJF-CNRS 5525, Faculty of Medicine of Grenoble
France

1. Introduction

Playing tennis or ski jumping in one's living room is fashionable. Being equipped with a self-regulated heating or light exposure system is convenient. Finding one's way in the middle of nowhere is reassuring. Smart devices have already flooded one's home, car, mobile phone, etc. Health Smart Home systems are not just trendy neither comfortable, they are necessary. Indeed, the worldwide population increases and ages. Moreover, lack of medical staff and suited infrastructures will become a real issue shortly. Aging in place should not be a request difficult to fulfill anymore. To meet this need, many studies have been led, these two last decades, on the development of biomedical devices which aim at improving the elderly quality of life, avoiding their entrance in dependence and thus in institution as much as possible (Cook *et al.* 2009, Chan *et al.* 2009). Telemonitoring is among the innovative technologies explored for the maintaining of elderly people at home. It consists in the follow-up of the subject behaviour, activities and more generally health state by means of ubiquitous smart sensors either placed in the environment (infrared or radar detectors, pressure sensors, video cameras, etc.) or embedded (GPS, accelerometers, etc.) (Chan *et al.* 2008).

This chapter discusses the ability to obtain reliable pervasive information at home from a network of localizing sensors allowing to follow the different activity-station at which a dependent (elderly or handicapped) person can be detected (Fouquet *et al.* 2009 a, b). The main idea is to watch the person at home in order to classify its activities of daily living, detect early its abnormal states or behaviour like perseveration, diagnosis and prolong its autonomy. Alzheimer or dementia related diseases are the first cause of autonomy loss and hence of entrance in institution (World Alzheimer report 2009). Detecting their onsets earlier is crucial for treatment effectiveness and to put the entry in dependence back. Following up the subject's sequence of locations may allow detecting spatial and temporal disorientation or abnormal behaviour. Has the subject got a clear goal or does he/she wander within his/her flat? Does he/she manage to perform daily living activities well? Questionnaires and scoring already exist (*e.g.* Katz basic Activities of daily living (ADL) scale, Katz *et al.*

1963, and the Lawton-Brody Instrumental Activities of Daily Living (IADL) scale, Lawton and Brody 1969). However, by fear of consequences or by shame, elderly people tend to lie to their doctor and do not admit their difficulties. Isolated consultations are not enough to detect this kind of symptoms whereas the monitoring of the subject in his/her owns environment may be a more accurate and reliable autonomy measurement tool. Perseveration may be a good indicator of autonomy loss beginnings (Miyoshi 2009, Joray *et al.* 2004, Sebastian *et al.* 2001, 2006).

The second section is a brief review of the Health Smart Home concept and techniques associated with. It includes the description of our platform of experiments located in Grenoble, France.

The data collection is described in the third section.

Then, in the fourth section, two approaches of perseveration modelling are introduced and a procedure to quantify their respective relevance is proposed. In the first approach, the succession of locations is interpreted as a sequence of coloured balls withdrawn from a generalized Pólya's urn. Each location corresponds to a given colour. Persistence is represented by adding balls of the last colour withdrawn. This model seems particularly pertinent for watching. In the second model, the succession of location is seen as a walk in a first order Markov chain. In this case, a location depends only on its predecessor. Persistence is interpreted as the probability to remain in the same activity-station. This approach has the advantage of being easily implemented.

The fifth section is dedicated to a location prediction procedure using a n-grams approach. It allows to determinate the order of the Markov chain which suits the most to predict the next location. Deviations from the predictions may be used as an indicator for alarm triggering.

Tools for decision-making are proposed in the sixth section.

Finally, numerical application of the processes introduced in the two last sections are presented in the seventh section and discussed in the eighth one.

2. Home Telehealth

2.1 Health Smart home concept

"Health Smart Home" (HIS in French) is a concept introduced during the last two decades (Cook *et al.* 2009, Chan *et al.* 2009), referring to the entrance of technologies dedicated to health into the household. These technologies are not limited to the monitoring of the inhabitant, they watch over him/her and act accordingly. The recent development of micro- and nanotechnologies allows the devising of ever more miniaturized sensors blended in the background, the clothes, everyday items, etc. These sensors acquire, in an automatic and unsupervised way, information about both the inhabitant and its surroundings. Three kinds of data may be distinguished: physiological (weight, skin temperature, blood pressure, cardiac and respiratory frequencies, blood glucose level, etc.), actimetric (posture, movements, walking trajectories) and environmental ones (ambient temperature, relative humidity of the air, noise level, luminosity, etc.). Recorded data are gathered and transmitted to a master computer which processes them while taking into account knowledge about the user. The software used must be able to fuse, analyse data from the sensor network, compare them to the user profile, update this profile according to what it learns from the data and trigger alarms if need be. All these actions must be fulfilled in a very short time to ensure the user's safety. Although both computers and data-processing techniques (data fusion, data-mining) are more and more efficient, we believe that a too large amount of data may be counterproductive. Consequently, the choice of the sensors (type, number, data format, etc.) and the sample frequency must be done strategically. Two kinds of approaches emerge. Some prefer multiplying sensors (video, audio, sophisticated, etc.) without being concerned about the cost, the energy consumption, the limited data storage (Chan *et al.* 2009). The others favour as far as possible the use of simple sensors dispatched in relevant places. Anonymous, binary sensors are cost-effective in terms of treatment as well as investment. More sophisticated sensors are used for more specific or accurate monitoring (physiological data, prevention of bedsore formation, etc.). However that may be, a preliminary classification work is made to quantify the relevance of the sensors chosen (Fleury *et al.* 2008, Rammal *et al.* 2008).

2.2 Smart sensors

Smart sensors field has moved with the development of Micro-Electro-Mechanical Systems (MEMS), telecommunication (internet, wireless network) and data-processing techniques (Sammarco *et al.* 2007, Huijsing 2008). Smart means that the sensor does not content itself making measurements, it also includes a pre-treatment of the data (quantifying the data quality, self-test, etc.). The miniaturization of the sensors allows their placement wherever in the subject surroundings: in article of clothing, devices, furniture, wall, doors, etc. They become nearly transparent for the user. This feature is very important to ensure the user's comfort but also to make him/her forget the telemonitoring system. Indeed, the elderly are very sensitive to a modification of their environment. Thinking continuously to the sensors may lead them to change voluntary or not their behaviour. It is well-known that by fear, some elderly are able to lie about their state of health to avoid a hospitalization or a supplementary cure. To understand the situation the user is in (context-awareness: Das and Roy 2008) without watching him/her directly, many kinds of sensor have been developed. Motion sensors like PIR detectors or RFID (Radio Frequency Identification) tags may be placed on the wall to follow-up the localization of the inhabitant. Sensors located on everyday items (*e.g. fridge, kettle, under the chairs, the bed...)* allow quantifying the interaction the user has with these objects. They may provide enough information to determine what task the user is performing. Embedded sensors (*e.g.* sewed within a cloth) may provide physiological (ECG, respiratory frequency, skin temperature) or actimetric data (posture, fall *via* 3D-axis accelerometer). In many cases, using only one kind of sensors is insufficient to ensure the safety of the subject that is why different kinds of sensor are usually combined. Concerning health, no rough estimate is bearable but minimizing uncertainty has a cost. Both technical and economical constraints (limited processing power, communication bandwidth, energy resources...) are added to the equation limiting the amount of sensors used. This optimization issue is the stake of classification which develops algorithms able to assess the relevance of the different sensors according to what is observed.

2.3 Data-processing tools

Data collected from the sensor network are gathered together and sent to a central where they are analysed. In case of missing/censored data, a reconstruction phase may be included to estimate them. Completed data are then fused. Data fusion consists in combining data extracted from different sources to obtain better information. For instance knowing that the subject is in the kitchen is not sufficient to claim that he/she is cooking. On the other hand, knowing that the subject is in the kitchen, the fridge door opened and the gas cooker switched on may be a good indicator of his/her activity. These preliminary steps are important for the reliability of the pervasive information. To detect abnormal behaviour, a "normal" profile of the user needs to be established. However, due to inherent inter-individual variability, it is not possible to create a general model. Therefore, the system has to be able to learn (*i.e.* deriving knowledge about) the user's behaviour from the data in an automatic and unsupervised way. Activity recognition is another expected feature of the system. It was very investigated the last ten years (Hong *et al.* 2009, Das and Roy 2009). To ensure the user's safety or to detect mild cognition impairment onsets, it may be useful to know if the daily living activities are well performed (occurrence, duration, accomplishment, etc.). Moreover, being able to predict the following activity could represent a real advantage. Deviations from the expected behaviour may become a supplementary tool revealing either a non-pathological change in the habits and then the necessity to update the user's profile or a suspicious change and then the necessity to trigger alarm.

2.4 The HIS and AILISA experimental platforms

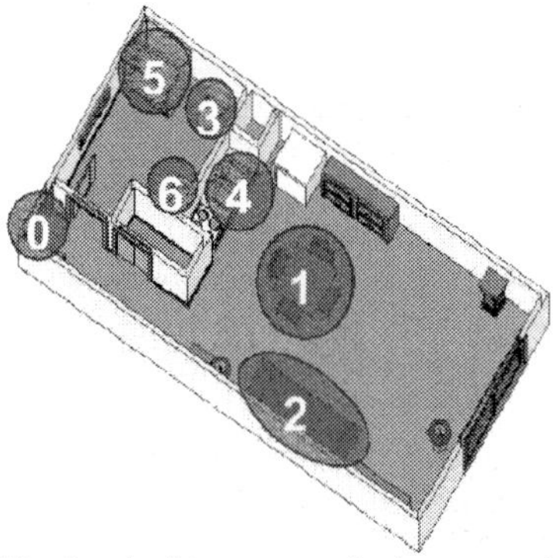

Fig. 1. Architecture of the experimental health smart home. Location sensors are placed at different places in the apartment, allowing the monitoring of individual's successive activity phases within his/her home environment: 0. Entry hall - 1. Living room - 2. Bedroom - 3. WC - 4. Kitchen - 5. Shower - 6. Washbasin.

Fig. 2. Infrared (arrows) for localizing dependent people in a health smart home

The HIS platform has been developed for a decade (Rialle *et al.* 2002; Demongeot *et al.* 2002; Virone *et al.* 2002, Le 2008). It is an experimental flat of 30m² located in the Faculty of Medicine of Grenoble, used for technological and clinical evaluation. At first sight, this flat looks common but on closer examination it turns out this flat abounds with sensors capturing all the day long measurements about the inhabitant (localization *via* PIR sensors, mobility), and his/her surroundings (temperature, hygrometry). Data are collected and transmitted *via* a controller area network (CAN) to a master PC placed in a dedicated room. Then, software analyzes automatically the data, compares them to the user's profile to detect abnormal trends and eventually trigger alarms. This system is passive *i.e.* it works automatically, without the inhabitant intervention. This feature is particularly important concerning the elderly who are not used to handle current technologies. Moreover, the use of anonymous sensors does not breach of the privacy of the followed person contrary to video cameras which must be used very carefully. Any private residence may be equipped with such a simple system. The HIS concept may be extended easily to other environments such as Hospital suites (LeBellego *et al.* 2006, Noury *et al.* 2008), office, public places, etc.

During the AILISA project (Noury 2005), the HIS platform was transposed in two real flats (Fig. 1) of the "Notre Dame" residence for the elderly in Grenoble (Fouquet *et al.* 2009 a, b). Passive IR sensors (DP8111X, ATRAL) were placed in each room (Fig.2) to follow-up automatically and continuously the localization of the inhabitant, an older woman, aged 80, within her own flat. Monitoring her localization in time provides a good approximation of her sequence of daily living activities. The aim of this experiment is to follow-up the sequence of daily routine of the inhabitant at home in order to detect a possible loss of autonomy or the emergence of a pathological behaviour such as perseveration. In particular, the present work focuses on the location modelling and prediction. Easy procedures are proposed to interpret surveillance at home data and to provide a perseveration index which may used to trigger alarms and counselling diagnosis search for cognition impairment (Miyoshi 2009, Joray *et al.* 2004, Sebastian *et al.* 2001, 2006). To reduce the complexity of the problem, a preliminary hypothesis is adopted. We only consider atomic *i.e.* elementary tasks. Therefore, we may identify a task and the room in which it is executed and talk about without distinction.

3. Experimental procedure

Recording time-stamped locations allows us to create a corpus for experiments. Timestamps are space separated numerals representing day of month, month, year, hour, minutes, seconds of the location captured (Table 1). The location itself is a code (Fig 1). The example cited in Table 1 means that the subject was in the kitchen the 3rd of August 2007 at 12:04'36". Data were collected during 10 months from the 22nd of March 2005 until the 24th of January 2006 and 6 months from the 18th of July 2007 to the 15th of January 2008. The corpus has been cut into 80% for learning model, 20% for testing it.

Day	Month	Year	Hour	Minute	Second	Activity-Station Code	Corresponding corpus line
03	*08*	*2007*	*12*	*04*	*36*	*4*	*03 08 2007 12 04 36 4*

Table 1. A time-stamped location and its corresponding translation into the corpus.

To make these data easier to handle, they were reformatted as followed. A line of the "new" corpus is a sequence of length 86,402 which represents a day as a series of location captured at each second and separated by a space. For instance, "*s 2 2 2 ... 2 2 3 3 3 ... 3 3 4 4 4 ... e*" means: since "*s*" the start of day, the subject was in the bedroom (2) where he/she spent x seconds (x is the number of successive 2), he/she passed in the toilet (3), then after y seconds (y is the number of successive 3), he/she went to the kitchen (4), etc. The end of day is represented by "*e*".

4. Persistence modelling approaches

The easiest way to analyse location data and extract from them preliminary trends in daily routine consists in the establishment of presence curves (Fig 3.)(Virone, 2009). To go on further, more sophisticated random processes techniques may be used. Classical time series techniques, like Box-Jenkins auto-regressive processes, have already been tested for modelling location succession (Das and Roy 2008, Virone *et al.* 2002, 2003 a, b, c). Among the various possible approaches for modelling the actimetric data, two methods have been selected. The first one is based on a generalized Pólya's urns scheme (Fouquet *et al.* 2009 b) in which the observed activity at time *t* depends on the whole past (since a reset supposed to be made at the beginning of each day). The second one concerns a first order Markov chain approach in which the dependency of the future of *t+1* lies only through the present time *t*. In both models, a persistence parameter is defined. To decide which method suits the most, a criteria based on the empirical mean E_i of remaining duration in a task *i* was proposed.

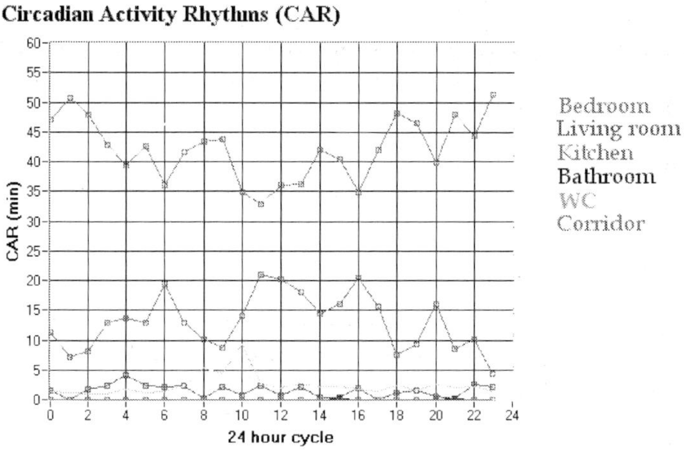

Fig. 3. Presence curves in different flat rooms

4.1 Pólya's urns model

Pólya's urns were used in 1923 by Eggenberger and Pólya to model the spread of contagious diseases. Since then, they were applied to various domains like climatology for the sequence of dry and wet days (Galloy *et al.* 1983). The success of the Pólya's urns scheme may be accounted by the fact that it is a visual mechanism easy to interpret in contrast with some abstract principles of probabilities (Kotz *et al.* 2000, Inoue and Aki 2001).

Regarding the scheme, a generalized Pólya's urn is an urn containing initially b_0 balls of N different colours split as follow: $a_0(i)$ balls of colour i, for i from 1 to N, with $b_0 = \sum_{i=1}^{N} a_0(i)$.

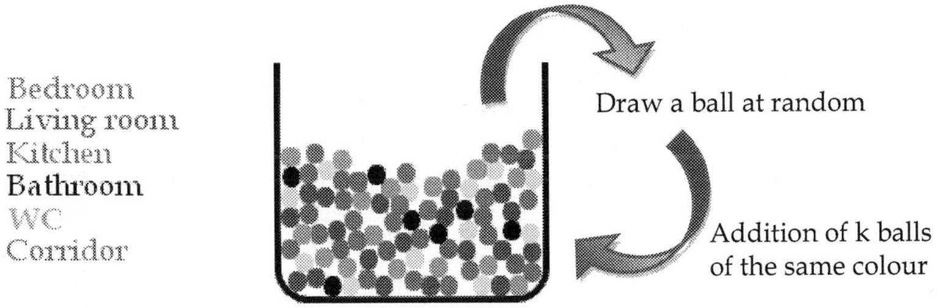

Fig 4. The generalized Pólya's urn scheme used for modelling location. Each colour represents a room of the flat. Persistence is taken into account by adding balls of the last colour drawn.

In our case, each colour corresponds to a location in the flat. At each step (*i.e.* each second here), a ball is withdrawn, its colour pointed out and the ball is put back in the urn along with additional balls of the same colour (Fig 4.). This last feature is precisely the way we chose to represent the persistence of an action/a location (Demongeot *et al.* 2008, Fouquet *et al.* 2009 a, b). In this approach, the persistence in task i parameter $\pi_i(t)$ is the ratio between the number $k_i(t)$ of balls of colour i added after the withdrawing of a ball of colour i at time t and the initial total number of balls b_0 unknown *a priori*.

$$\pi_i(t) = \frac{k_i(t)}{b_0} \tag{1}$$

In case of no pathologic change (sudden due to a fall or chronic due to the entrance in a neuro-degenerative disease), $k_i(t)$ does not depend explicitly on the time t, but only on the activity/station-code i. On top of that, we suppose that k_i does not depend on i either by considering that each room has the same role. Then, we are able to estimate π_i from the empirical frequencies $f_i(t)$ to get a ball of colour i at the $(t+1)^{th}$ draw and the number $x_i(t)$ of times where the ball of colour i has been drawn from the urn at time t (estimated in a series of days supposed to be independent).

$$\bar{\pi}_i(t) = \frac{f_i(0) - f_i(M)}{Mf_i(M) - x_i(M)} \tag{2}$$

where M is the total number of draws performed during a day *i.e.* 86,400 for a frequency sampling of one second.
Other assumptions on the $k_i(t)$ were further envisaged in (Fouquet *et al.* 2009 b).

To determine the domain of validity of this approach, we then estimate the expectation of the remaining duration in the task i denoted by E_i. The simplest estimator $\overline{E_{\iota,1}}$ is obtained by considering the empirical (*i.e.* on observed days) mean of the remaining duration in a day and given by:

$$\overline{E_{\iota,1}} = \frac{1}{M+1}\sum_{t=0}^{M} z_i(t) \tag{3}$$

where $z_i(t)$ is the length of the sequence of "drawing a ball of colour i" (possibly 0) since a draw at time t of a ball of colour i.
A more accurate estimator $\overline{E_{\iota,2}}$ may be calculated by approximating the probability $c_{i,m}(t)$ to have m consecutive draw(s) of a ball i from the tth draw. It is given by:

$$\overline{E_{\iota,2}} = \frac{\sum_{t=0}^{M}\sum_{m=0}^{M} m[(1-f_i(t+m+1))\prod_{j=0}^{m} f_i(t+j)]}{M+1} \quad (4)$$

Then the 95%-confidence interval of $\overline{E_{\iota,2}}$ may be calculated based on the f_i's one which is:

$$\left[f_i \pm 1.96\sqrt{\frac{f_i(1-f_i)}{M}}\right] \quad (5)$$

The null hypothesis H_0: "the persistence model is a Pólya's urn model" is rejected, if $\overline{E_{\iota,2}}$ does not belong to its interval. Otherwise, this model could be used to represent the persistence in task.

4.2 First order Markov chain model

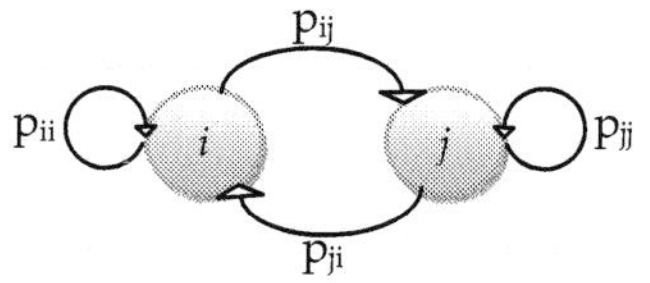

Fig. 5. The Markov chain obtained for two rooms i, j

In a first order Markov chain approach, the location at time t depends only on its predecessor (Fouquet *et al.* 2009 a). Each location is represented by a node (N nodes altogether) fitted with the probabilities of transitions from it to another one as illustrated in Fig 5. The succession of locations is seen as a walk in the Markov chain. Let us denote by p_{ij} the probability, supposed to be constant (*i.e.* each room has the same role), to draw a ball of colour j after a ball of colour i. Then, p_{ii} may be seen as the persistence in task i parameter and easily estimated by the corresponding empirical frequency f_{ii}. The probability p_j to draw a ball of colour j is obtained by adding the probabilities of transitions encountered on each arrow which points to the node j *i.e.*by considering all the ways leading to j:

$$p_j = \sum_{i=1}^{N} p_{ij} \quad (6)$$

In this approach, the variable z_i above-mentioned is easy to interpret:

$$P(z_i = 0) = 1 - p_i \quad (7)$$

$$P(z_i = l) = p_i(1-p_i){p_{ii}}^{l-1} \quad \text{for } l \in [\![1, N]\!] \quad (8)$$

Thus, E_i may be estimated by:

$$\overline{E_{i,3}} = \sum_{l=0}^{M} \frac{l+1}{2} f_i(1-f_i) f_{ii}^{\ l-1} \quad (9)$$

The 95%-confidence interval of $\overline{E_{i,3}}$ may be calculated based on the f_i's and f_{ii}'s ones which are respectively:

$$\left[f_i \pm 1.96\sqrt{\frac{f_i(1-f_i)}{M}}\right]; \quad \left[f_{ii} \pm 1.96\sqrt{\frac{f_{ii}(1-f_{ii})}{M}}\right] \quad (10)$$

The null-hypothesis H_0: "the persistence model is a first order Markov chain model" is rejected, if $\overline{E_{i,3}}$ does not belong to its interval. Otherwise, this model could be used to represent the persistence in task.

At the end of the hypothesis testing, in case of acceptation for both models, the Markovian model will be chosen because of its simplicity and its handiness. If both tests conclude to the rejection of the null-hypothesis, the model having the closest distance between $\overline{E_{i,1}}$ and the confidence interval of $\overline{E_{i,j}}$ (j=2, 3) will be retained.

We will focus now on the description of the first order Markov chain model and its generalization by the *n*-grams approach proposed. The latter aims at providing the best order of the Markov chain *i.e.* the rank of the last data on the historic of locations needed to determine the present location.

	Pólya's urns model	**First order Markov chain model**
Estimator of the persistence in the ith task parameter	$\bar{\pi}_i(t) = \dfrac{f_i(0) - f_i(M)}{Mf_i(M) - x_i(M)}$	f_{ii}
Estimator of the ith task remaining duration	$\overline{E_{i,1}} = \dfrac{1}{M+1}\sum_{t=0}^{M} z_i(t)$ $\overline{E_{i,2}} = \dfrac{\sum_{t=0}^{M}\sum_{m=0}^{M} m[(1-f_i(t+m+1))\prod_{j=0}^{m} f_i(t+j)]}{M+1}$	$\overline{E_{i,3}} = \sum_{l=0}^{M} \dfrac{l+1}{2} f_i(1-f_i) f_{ii}^{\ l-1}$

Table 2. A time-stamped location and its corresponding translation into the corpus.

5. Statistical location prediction: *n*-grams model

5.1 Fundamental basement

Predicting one's activities during the day seems *a priori* a very difficult task. One may perform the same activities every day but not necessary in the same order. One's daily routine may also be disturbed by tiredness, visits, etc. Actually, as well as the human being's processes and physiological variables (blood sugar level, hormone level, temperature, cardiac rhythm, muscular strength, etc.) are regulated by biological internal rhythms and are well-organized in time, activities of daily living also follow periodical variations adapted, in this case, to both ones biological and social rhythms (Reinberg 1998, Virone *et al.* 2002, Demongeot *et al.* 2002). The elderly are less sensitive to the clocks of society and their way of living is slower. Hence, the observation of their daily routine is easier and is less subjected to the unexpected. Indeed, with aging, the everyday procedure becomes more and more stable allowing the establishment of an activity prediction procedure. Important and unexpected deviations from the behaviour expected may be a good indicator for triggering alarms. The development of such a procedure is difficult because of individual variability. Each person has his/her own habits and even for the same person, depending the day of the week, the season, etc. or also fatigue state, the sequence of activities may be different. In the speech recognition domain, this problem is encountered for word prediction. Statistical approaches based on *n*-grams theory are often used to overcome these kinds of difficulty (Shannon 1948, Rosenfeld 1996, Fouquet 2004). The following experiment aims at applying the *n*-grams theory to location modelling problem. The model used is a generalisation of the first order Markov chain model.

5.2 *N*-grams model

For the location prediction, a statistical method has been implemented to predict the next location on the basis of the location history (Reithinger & Maier 1995). Currently, *n*-grams location probabilities are used to compute the most likely follow-up location. To predict the ith location a_i, the *n-1* previously uttered locations are used and the most probable location is determined by computing:

$$a_i = \arg\max_a P(a | a_{i-1}, a_{i-2}, \dots, a_{i-n+1}) \quad (11)$$

To estimate this probability the standard estimations using relative frequency techniques are used. Otherwise, our corpus is a real-collection and, like in many real-situations, it was not possible to collect a large amount of data to properly estimate the statistics. This implies that it is not reasonable to use classical smoothing techniques. We need a solution for the two following problems:

1) unexpected input: the location model based on *n*-grams location sequences cannot be used in case unexpected input occurs,

2) lack of training data: the *n*-grams model predicts several locations with the same probability.

The treatment of these cases consists in using the *(n-1)*-grams model, recursively.

6. Decision-making tool

If and as soon as deviations from the expected behavior occurred, a smart home system must 'decide' to alert caregivers. The decision-making process is based on the information extracted from a Bayesian network. The advantage of using such a representation is that it allows us to combine both *a priori* knowledge (e.g. from the doctor) and collected data. Besides, it was recently used to mimic the brain decision-making process (Knill *et al.* 2004,

Koerding *et al.* 2004) in which both *a priori* distribution (knowledge about the environment) and observed random variables (metrological sensors) are taken into account. Another application of such a data fusion may be encounter in postural control correction in which the action consists in preventing the subject in case of bad position of the body. It is generally agreed that maintaining an upright stance or seated posture involves the integration (fusion) of sensory information from multiple natural sensors including visual (variable X_1), somatosensory (variable X_2) and vestibular (variable X_3) systems. The main idea is to provide supplementary sensory information to complete/improve the existing one. Along this line, innovative health technologies have been recently developed for pressure sores or fall prevention in older and/or disabled adults by the means of artificial sensors placed beneath the buttock or the feet, *via* an alternative sensory modality (electrotactile stimulation of the tongue, variable X_4) (Vuillerme *et al.* 2007 a, b). At this point, an effective fusion of natural (variables X_1, X_2 and X_3) and artificial - more reliable and accurate than the natural one's - sensory information (variable X_4) is crucial to enable individuals with spinal cord injuries, or with somatosensory loss in the feet from diabetic peripheral neuropathy to become aware of a localized excess of pressure at the skin/seat interface and/or postural orientation and thus to make adaptive postural corrections.

In this kind of multisensory contexts, by denoting X_i (for i from 1 to n) the n observed random variables, the conditional probability to execute a task A knowing the values of X_i's is given by:

$$P(A|\cap_{i=1}^{n}\{X_i = k_i\}) = P(A \cap \cap_{i=1}^{n}\{X_i = k_i\})/P(\cap_{i=1}^{n}\{X_i = k_i\}) \qquad (12)$$

The calculation of such a probability requires the ability to estimate joint probabilities of the n observations. The Lancaster-Zentgraf estimator defined in the seventies was proposed to estimate of the joint probability of order n knowing the marginal and joint probabilities of order 2 as follow:

$$P_{Lan}\left(\bigcap_{i=1}^{n} A_i\right) = \sum_{\substack{i,j,k_1,\dots,k_{n-2}\in\{1,n\},\\ i\neq j\neq k_1\neq\dots\neq k_{n-2}}} \left(P(A_i \cap A_j)P(A_{k_1})\dots P(A_{k_{n-2}})\right) - \binom{n}{2}\prod_{i=1}^{n} P(A_i)$$

where $A_i = \{X_i = k_i\}$. (13)

For n=3, the equation becomes:

$$P_{Lan}(A \cap B \cap C) = P(A \cap B)P(C) + P(A \cap C)P(B) + P(B \cap C)P(A) - 2P(A)P(B)P(C) \qquad (14)$$

However, it happens that this approach provides negative estimates (*e.g.* A, B, C disjoint). To alleviate this issue, we introduce a new joint probabilities estimation P_{New} based on the local equipartition of the amount of uncertainty (in a local maximal entropy approach) (Demongeot *et al.* 2008). The proposed formula is established to deal with dependencies characterized by strong incompatibilities, circumstances not well taken into account by the Lancaster-Zentgraf formula:

$$P_{New}\left(\bigcap_{i=1}^{n} A_i\right) = \frac{1}{n}\sum_{j=1}^{n} P_{New}\left(\bigcap_{i\neq j} A_i\right)P(A_j) \qquad \text{for } n>2 \qquad (15)$$

For the intersection of any 3 events from a set of n events, this equation becomes:

$$P_{New}(A \cap B \cap C) = \frac{1}{3}\big(P(A \cap B)P(C) + P(A \cap C)P(B) + P(B \cap C)P(A)\big) \tag{16}$$

In practice, the calculation of P_{New} is done recursively from the calculation of P_{New} on the triplets of events involved in $\bigcap_{i=1}^{n} A_i$, that involves as for the Lancaster-Zentgraf estimator the knowledge of the marginal and order 2 intersection probabilities. Finally, we have:

$$P_{New}\left(\bigcap_{i=1}^{n} A_i\right) = \frac{1}{\binom{n}{2}} \sum_{i<j} P(A_i \cap A_j) \prod_{k \neq i,j} P(A_k) \tag{17}$$

7. Results

7.1 Pólya *vs.* Markov

The corpus of experiment was used to calculate the empirical means $\overline{E_{i,j}}$ (j=1,2,3) of remaining duration in task i. To achieve this calculation, Tables 3 and 4 show the empirical frequencies f_i and f_{ij} respectively, calculated from the 20% learning part of the corpus. As above-mentioned, M is the number of locations recorded during a day at a sampling frequency of 1 second (*i.e.* $M = 60 \times 60 \times 24 = 86400$).
The calculation of $\overline{E_{i,1}}$ consists in counting for each day the time remaining in task i divided by the number of times observed (which is equal to $M+1$ if the observation starts from 0 to M). It expresses persistence in task i, but it is not equal to the mean of past time in i (it should be half the preceding one). One can now distinguish two particular cases:
- If i was never observed: $\overline{E_{i,1}} = 0$
- If i was always observed: $\overline{E_{i,1}} = \frac{\frac{(M+1)(M+2)}{2}}{M+1} = \frac{M+1}{2} = 43{,}200.50$

For the other cases, further works have to be done now to calculate $\overline{E_{i,1}}$.
In the other hand, E_i can be easily estimated by the Markovian estimator $\overline{E_{i,3}}$ which values are shown in Table 5 for each location i.

i	0	1	2	3	4	5	6	9	e
f_i (%)	10.27	21.2	53.18	1.29	6.13	1.67	5.13	1.12	0.01

Table 3. Empirical frequencies f_i to draw a ball of colour i (%). Note that the *9* is an error.

j \ i	0	1	2	3	4	5	6	9	e
0	99.64	0.12	0.04	0.01	0.03	0.10	0.06	0.00	0.00
1	0.07	97.48	0.19	0.01	2.20	0.03	0.02	0.00	0.00
2	0.01	0.08	99.89	0.00	0.00	0.01	0.00	0.00	0.00
3	0.17	0.14	0.08	81.29	0.06	3.00	15.26	0.00	0.00
4	0.05	7.62	0.03	0.01	92.24	0.03	0.01	0.01	0.00
5	0.56	0.54	0.054	1.84	0.13	86.50	9.10	0.79	0.00
6	0.07	0.02	0.02	4.02	0.01	2.81	90.63	2.42	0.00
9	0.06	0.04	0.01	0.00	0.07	1.51	10.74	87.57	0.00
s	0.19	0.58	9.24	0.19	0.13	0.06	0.45	0.26	88.89

Table 4. Empirical frequencies f_{ij} to draw a ball of colour j after a ball of colour i (%).

The confidence intervals are shown in Table 6 for f_i and f_{ij}, respectively. They are necessary to derive the confidence interval of each estimator $\overline{E_{\iota,J}}$ (j=1, 2, 3) and to quantify their relevance.

i	*0*	*1*	*2*	3	*4*	*5*	*6*	*9*	*e*
$\overline{E_{\iota,3}}$	3,596.89	133.15	102,829.88	0.20	4.97	0.48	2.91	0.38	0.01

Table 5. Estimator $\overline{E_{\iota,3}}$ of the ith task remaining duration using the first order Markov Model.

	i	*0*	*1*	*2*	*3*	*4*	*5*	*6*	*9*	*e*
f_i	a	0.101	0.209	0.528	0.012	0.060	0.016	0.050	0.010	0.3e-4
	b	0.105	0.215	0.535	0.014	0.063	0.018	0.053	0.012	1.6e-4
f_{ii}	a	0.996	0.974	0.9987	0.812	0.921	0.864	0.905	0.875	0.888
	b	0.997	0.975	0.9990	0.814	0.923	0.866	0.907	0.877	0.890

Table 6. Confidence interval [a,b] for f_i and f_{ii} respectively.

7.2 Location prediction

	Global corpus (2,011,554)		Sunday (300,105)		Monday (267,983)	
n	Number of correct predictions	Good prediction rate	Number of correct predictions	Good prediction rate	Number of correct predictions	Good prediction rate
1	983,209	48.88 %	154,846	51.60%	126,858	47.34%
2	1,958,535	97.36 %	291,248	97.05%	261,253	97.49%
3	1,958,535	97.36 %	291,250	97.05%	261,254	97.49%
4	1,958,524	97.36 %	291,244	97.05%	261,252	97.49%
5	1,958,481	97.36 %	291,258	97.05%	261,222	97.48%
6	1,958,380	97.36 %	291,190	97.03%	261,212	97.47%
7	1,958,297	97.35 %	291,192	97.03%	261,181	97.46%
8	1,958,221	97.35 %	291,143	97.01%	261,148	97.45%
9	1,958,062	97.34 %	291,121	97.01%	261,113	97.44%
10	1,957,773	97.33 %	291,043	96.98%	261,057	97.42%

	Tuesday (289,876)		Wednesday (427,006)		Thursday (232,471)	
n	Number of correct predictions	Good prediction rate	Number of correct predictions	Good prediction rate	Number of correct predictions	Good prediction rate
1	123,530	42.61%	197,499	46.25%	136,005	58.50%
2	284,163	98.03%	416,983	97.65%	226,140	97.28%
3	284,162	98.03%	416,984	97.65%	226,129	97.27%
4	284,156	98.03%	416,980	97.65%	226,106	97.26%
5	284,129	98.02%	416,953	97.65%	226,099	97.26%
6	284,105	98.01%	416,909	97.64%	226,085	97.25%
7	284,089	98.00%	416,889	97.63%	226,070	97.25%
8	284,057	97.99%	416,841	97.62%	226,047	97.24%

9	284,002	97.97%	416,759	97.60%	226,024	97.23%
10	283,950	97.96%	416,687	97.58%	225,961	97.20%

Table 7a. Prediction results for the global corpus in one hand, for each day of the week in the other hand (the number to predict is in parenthesis).

	Friday (260,158)		Saturday (199,767)	
n	Number of correct predictions	Good prediction rate	Number of correct predictions	Good prediction rate
1	137,179	52.73%	91,473	45.79%
2	252,605	97.10%	194,207	97.22%
3	252,597	97.09%	194,207	97.22%
4	252,596	97.09%	194,200	97.21%
5	252,594	97.09%	194,194	97.21%
6	252,550	97.08%	194,192	97.21%
7	252,519	97.06%	194,174	97.20%
8	252,445	97.04%	194,126	97.18%
9	252,386	97.01%	194,121	97.17%
10	252,325	96.99%	194,085	97.16%

Table 7b. Prediction results for the global corpus in one hand, for each day of the week in the other hand. The number above each day represents the number to predict.

A first test (Table 7a.) was made on the whole corpus without time distinction (day of week, day of month, month, hour of journey, etc.). It turns out that the best prediction is reached with n=2. Indeed, nearly the same performance is obtained with n>2 (with two digits after the decimal point). Accurate performance decreases while n increases. Hence, n does not need to be bigger than 2, suggesting that the last second location is sufficient to predict the next one. Finally, this first result seems to indicate that watching too far in the past is not a good way to predict the future location of a person. It is in favour of a first order Markov approach.

Even if results from the global corpus are promising, better accuracy may be obtained by distinguishing day of the week (Table 7a, b.). Indeed, activity of an elderly person during the weekend is likely to differ from the rest of the week (time spent with family, Church service, etc.). Concerning household chores, the flat is not cleaned up from top to bottom every day.

On Sunday, the best prediction rate corresponds with n=5 which means that the last four locations are needed to predict the next one). On Monday and Wednesday, it is necessary to date back to the last three ones two ones. Finally on Tuesday, Thursday and Friday, only the previous location is needed to have a good prediction rate.

In both cases, n=2 provides the best prediction rate with two digits after the decimal point. Results of good prediction vary from 97.05% on Sunday up to 98.03% on Tuesday.

8. Discussion and Conclusion

The monitoring of elderly people in their home by location sensors provides the recording of their walk trajectories within their own flat giving some insights into their daily routine. These space/time data are then used to establish an individual specific profile concerning the time serie which corresponds to the successive locations of the person. The detection of large deviations from the profile may be used to trigger alarms at the level of the family (to make the family member more careful) or at the level of the Hospital at Home service (for an emergency). The present article proposes two different mathematical approaches for location modelling. The first calculations point out the simplicity of the Markovian model. This feature makes it more cost-effective in terms of calculation and response time and thus more interesting for implementation. Further works are necessary to explore the Pólya's urns model deeper. If the choice is available, however, priority will be given to the other. The Pólya's urns model will only be used if the Markovian one is no longer available.
For the validation of each model, a procedure based on the calculation of the mean of remaining duration in task is proposed. At this point, it is important to mention that this parameter differs from the mean of the time passed the task considered (it should be by a factor of 0.5). For deciding between the two methods proposed, further works should be performed using the statistics equal to the empirical mean E of a task remaining duration and their confidence interval.

The n-grams model used to predict the next location is also in favour of a first order Markov chain approach. Other models need to be improved more deeply. Indeed, first results offer promising results with n=2 and a degradation of performances with n increasing up to 10. For the confirmation of this trend, experiment should be applied for n=*60*, watching for the whole last minute in order to predict the *60th* second. For the time being, the best model offers up to 98.03% of good prediction location, considering only the last second of location *i.e.* using the first order Markov chain model, but distinguishing days of week. Indeed, taking day of the week into account offers better performance (97.36% by working directly on the whole corpus). Performance seems to differ for each day of week. This factor of variability should be taken into account when designing a system using a location model. To go further, future experiments may be done for other refinements among which providing the possibility of distinguishing the time in the day to take into account the circadian rhythm (Virone *et al.* 2002, Cerny *et al.* 2009). Each day of the month could affect activity: some days, as *1st* of the month for example, are particular. The comparison between each month may show seasonal effects, and so on (Fouquet *et al.* 2009 b). It could then be interesting to develop a new model with a continuum approach considering estimations (interpolation) between data observed.

Our procedure may be extended later to more complex task (named composite task in Das *et al.* 2008) recognition by means of a multi-sensors network. The multidimensional data collection obtained with such a system will be fused and mined using the data processing techniques above introduced.
Concerning the detection of progressive stereotyped behaviour for the early diagnosis of neuro-degenerative diseases, actimetric data may be completed by physiological ones to quantify the state of anxiety of the subject particularly at nightfall or a possible shift in his/her activity circadian rhythm which are behaviour commonly observed in the first stages of these pathologies (Monk 2005, Hofman and Swaab 2006).

9. Acknowledgements

The data were recorded by the AFIRM Team from TIMC-IMAG Laboratory and RBI during the AILISA project (Noury 2005), supported by the French RNTS health network since 2003 within the framework of the "Institut de la Longévité" (n°03B651-9).

10. References

Cerny, M. & Penhaker, M. (2009). Circadian rhythm monitoring in HomeCare Systems. *Proceedings of IFMBE, 13th International Conference on Biomedical Engineering*, pp. 950–953, ISBN: 978-3-540-92841-6, Rep. of Singapore, Dec. 2008, Springer, Singapore.

Chan, M.; Estève, D.; Escriba, C. & Campo, E. (2008). A review of smart homes: Present state and future challenges. *Computer Methods and Programs in Biomedicine*, Vol. 91, No. 1, Feb. 2008, pp. 55-81, ISSN: 0169-2607.

Chan, M.; Campo, E.; Estève, D. & Fourniols, J.-Y. (2009). Smart homes - current features and future perspectives. *Maturitas*, Vol. 64, No. 2, Sept. 2009, pp. 90-97, ISSN: 0378-5122.

Cook, D.; Augusto, J. & Jakkula, V. (2009). Ambient intelligence: Technologies, applications, and opportunities. *Pervasive and Mobile Computing*, Vol. 5, No. 4, Apr. 2009, pp. 277–298, ISSN: 1574-1192.

Das, S. & Roy, N. (2008). Learning, prediction and mediation of context uncertainty in smart pervasive environments. *Proceedings of the OTM Confederated International Workshops and Posters on, On the Move to Meaningful Internet Systems*, pp. 820–829, ISBN: 978-3-540-88874-1, Mexico, Nov. 2008, Springer, Mexico City.

Demongeot, J.; Virone, G.; Duchêne, F.; Benchetrit, G.; Hervé, T.; Noury, N. & Rialle, V. (2002). Multi-sensors acquisition, data fusion, knowledge mining and alarm triggering in health smart homes for elderly people. *Comptes-Rendus Biologies*, Vol. 325, No. 6, June 2002, pp. 673–682, ISSN: 1631-0691.

Demongeot, J.; Noury, N. & Vuillerme, N. (2008). Data fusion for analysis of persistence in pervasive actimetry of elderly people at home, and the notion of biological age. *Proceedings of International Conference on Complex, Intelligent and Software Intensive Systems CISIS'08*, pp. 589–594, ISBN: 978-0-7695-3109-0, Spain, March 2008, IEEE Xplore, Barcelona.

Fleury, A.; Vacher, M.; Glasson, H.; Serignat, J. & Noury, N. (2008). Data Fusion in Health Smart Home: Preliminary Individual Evaluation of Two Families of Sensors. *Proceedings of 6th International Conference of the International Society of Gerontechnology*, pp. 135-139, Italy, June 2008, Pisa.

Fouquet, Y. (2004). Modélisation des attentes en dialogue oral. Ph.D. Dissertation. UJF, Grenoble.

Fouquet, Y.; Demongeot, J. & Vuillerme, N. (2009). Pervasive informatics and persistent actimetric information in health smart homes: From language model to location model. *Proceedings of International Conference on Complex, Intelligent and Software Intensive Systems CISIS '09*, pp. 935–942, ISBN: 978-1-4244-3569-2, Japan, March 2009, IEEE Explore, Fukuoka.

Fouquet, Y.; Vuillerme, N. & Demongeot, J. (2009). Pervasive informatics and persistent actimetric information in health smart homes, In: *Ambient Assistive Health and Wellness Management in the Heart of the City*, pp. 108-116, Springer, ISBN: 978-3-642-02867-0.

Galloy, E.; Le Breton, A. & Martin, S. (1983). A model for weather cycles based on daily rainfall occurrence, *Lecture Notes in Biomathematics*, Vol. 49, pp. 303–318.

8

Smart home with healthcare technologies for community-dwelling older adults

Machiko R. Tomita, Ph.D., Linda S. Russ, Ph.D., OTR/L,
Ramalingam Sridhar, Ph.D., Bruce J. Naughton, MD.
University at Buffalo, State University of New York
Buffalo, New York, U. S. A.

1. Introduction

Traditional smart home (SH) technology combined with new monitoring technologies designed to prevent falls and detect health status changes is the current direction for SH development. The combination of conventional SH technology with healthcare technology is likely to be an effective approach to preventing nursing home placement for frail older adults.

Barriers to adopting or implementing monitoring technology include concerns about privacy and reluctance of older adults to accept the technology. Acceptance of the monitoring technology depends not only on the older persons' health and functional status but also upon the perception of need and personal goals. Since great variability of health and disability conditions exists among community-dwelling older adults, their purposes of using SH with healthcare technologies may also be diverse depending on their levels of physical and cognitive functions and severity of illnesses. Older adults and their informal caregivers are likely to insist on choice of technologies suitable for their goals. Health care providers need to be familiar with the available technology.

Home-based technologies allow older persons to adapt to changing physical and cognitive function and preserve living in a familiar environment within an established social network. Home-based technologies may be most effective, if they also promote healthy behaviour and lifestyles for continued community living.

Focusing on the situation in the United States, this chapter includes: (1) a description of societal needs for the technology, (2) a summary of current SH status, (3) a review of older adults' perceptions of SH, healthcare technologies and their effectiveness, (4) caregivers' perspectives of the technology, and (5) defining the role of SH with health care technologies in the Model toward Optimal Management for Independent through Technological Adoption (Tomita et al., 2009). The model expands on behavioural medicine to promote healthy behaviours in community-dwelling older adults.

2. Societal needs for smart home and healthcare technologies in the United States

The United States is experiencing a rapid increase in its aged population. An aging population is associated with an increase in the incidence and prevalence of chronic diseases and an increase in health care expenditure (Administration on Aging, 2008). In 2000, the total Medicare expenditure was $239.5 billion; 10 years later, in 2010, expenditures are expected to rise to $519 billion (Foster, 2008). As 77 million baby boomers (born in 1946-1964) begin to reach age 65 in 2011, it is forecasted that Medicare expenditures will explode.
Aging in place has been proposed as one method to reduce cost and maintain quality of life for the aging population. The concept is to support older adults in the environment of their choice in lieu of institutionalisation or nursing home placement. To make this supportive healthcare possible, medical/health monitoring devices and e-health technology have been developed. Use of these technologies in a home setting is still at its dawn, but this method appears to be one of the most promising approaches to facilitate independent living in community-dwelling older adults when the use of the Internet is rapidly increasing. When a health monitoring system is imbedded in a smart home, older people can live in their home while receiving medical support. This idea has been addressed by Stefanov, Bien, and Bang (2004) of the health monitoring subsystem as an important component of SH.
Although SH with health monitoring system appears to be among the most promising methods to facilitate continued community dwelling among older adults, there are many unanswered questions. A world-wide study of longevity tells us that "de-convenience" is important to increase the amount of daily activities for a healthy and longer life (Pinkus, 2009). A healthy life requires physical activity and human connections to facilitate a healthy lifestyle. The convenience of remote control for televisions may compromise a healthy lifestyle. SH technology faces the challenge of convenience and the technology may reduce rather than promote human contact. Reliance on SH technology needs to balance convenience with the need to foster healthy behaviour and social interaction.

3. Current smart home with healthcare technologies status in the United States

In this section, a current status of SH with healthcare technologies for community-dwelling older adults in the United States is summarized in several categories: laboratory setting, prototype of SM, SH in use, and retrofitted SH. SH technologies in nursing homes are excluded in the summary.

3.1 Laboratory setting

SH with healthcare technology for older adults can be found most commonly in a laboratory setting at a university which utilizes an infrastructure equipped with sensors, panels, cameras, etc. The purpose is usually to research older adults' behaviour in a home environment.
The Smart Medical Home at University of Rochester's Center for Future Health at the University of Rochester, NY, is one example. The five-room house has infrared sensors, computers, biosensors, and video cameras. Some of the research involves a medication advisor who can converse with a research subject and advise on medication management

and dietary adherence, provide memory assistance, and assist with Smart Bandage. Smart Bandage seeks to decrease the burdens of chronic wound care at home. (Center for Future Health, 2005).

When this smart home is applied in a real setting, the integrated technologies may allow residents to maintain health, detect the onset of disease, and manage disease. Future applications include gait monitoring, and observation of behaviour and sleep, and adherence. The Smart Medical Home was designed for adults of all ages, but it is not meant for actual habitation.

3.2 Prototype of smart house

Gator Tech Smart House at the University of Florida-Gainesville's Mobile and Pervasive Computing Laboratory is a comprehensive laboratory-house created to assist older adults in maximizing independence and maintaining higher quality of life (Helal et al., 2005). The entire house is equipped with (a) smart cameras for motion detection, image processing, and control of other in-home devices, (b) smart blinds that automatically close to block sunlight when the air conditioner is on, (c) ultrasonic location tracking transceivers that are installed on the ceiling corners of each room to detect movement, location, and orientation of the resident, (d) smart floor that uses pressure sensors embedded into each tile to detect falls and reports to emergency services, and (f) smart displays for entertainment media and information residents can follow from room to room. Additional features include a smart mailbox that senses and notifies the arrival of mail, a smart front door that identifies residents, using a radio-frequency identification tag among others, a smart bed that monitors sleeping patterns, a smart mirror that displays important messages or reminders such as when to take medication, and a smart bathroom that includes a toilet paper dispenser, a flush detector, and a water temperature regulating shower. The Gator Tech Smart House is adding healthcare technologies to assist diabetes management (Helal, Schmalz, & Cook, 2009).

Similarly, a SH apartment called CASAS in the Washington State University has seventy motion detectors on doors, a regenerator, a microwave oven, under the sink, a bed, etc. to train a computer to monitor activities of daily living of people who have dementia, and when urgent situations arise, alert caregivers. The sensors and computer hardware were mostly purchased off-the-shelf. But other technologies such as software programming, pattern recognition, and artificial intelligence are being developed (KUOW, 2008).

Both the Gator Home and CASAS have integrated health monitoring software components within a comprehensive monitoring system; this is called the smart home-based health platform. For this, they took an approach to probabilistically identify activities in a SH from sensor data, while the activity is being performed, using a hidden Markov model. Data from the comprehensive monitoring system supported development of a mathematical model to describe sensor patterns for a given activity and for change in activity. The technology monitors diabetic patients' diet and exercise adherence, in addition to home-activities (Helal, Schmalz, & Cook, 2009). Neither Gator House, nor CASAS is designed for actual habituation.

Research continues with these prototype homes. A two-story single-family house called Aware Home by Georgia Institute of Technology (Atlanta, GA) is a living laboratory house designed primarily to assist an older adult with cognitive impairment. (Do, 2008). For example, there is a capture system on the kitchen countertop. If a person forgets how many

cups of flour to put in a mixing bowl, a wall display shows visual snapshots arranged as a series of panels for him/her to touch for review of activities. A similar system can be used to support safe and complete medication adherence (Georgia Institute of Technology, 2009).
While most of the technology in Aware Home is for older adults with cognitive deficit, healthcare technology was also developed for diabetes management in individuals without cognitive impairment. Mobile Access to Health Information is used to record readings through verbal descriptions and pictures. It uses a mobile phone to which a glucose meter can be connected via Bluetooth. This captures the impact of recorded activities on blood sugar and is available on password-protected websites. Health technology does not have to be home-bound.

3.3 Smart homes in use

Recently, advanced SH technologies have been implemented in actual community settings, apartment complexes, and retirement housing units. A smart home in Vinson Hall Retirement Community in Missouri is dedicated to serving former U.S. military officers and their families. It includes different coloured lighting to help people with vision impairments, a contiguous channelling kitchen where residents can move heavy objects without lifting, and a garbage can that opens with a wave of the hand. In this home, technology is designed to be less obvious and to feel as though it is part of a normal home. For example, grab bars in the bathroom look like upscale towel bars. (Crawford, 2009).
Eskaton, Ltd. created the National Demonstration Home in California with a range of technologies. The combination of SH and healthcare technologies includes a home monitoring system, Internet and teleconferencing abilities, brain fitness activities, and a wellness monitoring station that transmits daily health reports to caregivers. This home utilizes universal design features including a no-step entry, reduced height cabinetry, and an accessible flow to the floor plan. The universal design features cost an additional $4,000 - $8,000 for the 1,600-square-foot floor plan; residents pay an extra $100 to $150 a month to access the technology (Crawford, 2009).
The University of Missouri-Columbia integrated sensor networks into privately owned apartments called TigerPlace II. In the apartments, a health monitoring system detects changes in behaviour and physical activity, including walking and sleeping pattern without compromising privacy. Small sensors over the toilet, shower, and doorways detect residents' movements. Pneumatic tubes under the mattress and a chair measure weight. The assumption is early identification of older adults' behaviour changes can prompt healthcare interventions to delay or prevent serious health events. In addition, web-based information is available to healthcare providers (ANI, 2008). This system has been installed in 22 units and currently is in use (Medical New Today, 2009). This system proved effective for at least one resident. The resident's caregiver and researchers at the University of Missouri-Columbia identified patterns in the data that signalled exacerbation of congestive heart failure. Early intervention reduced severity of the episodes.
The most recent development in SH in the United Sates can be found in the McKIZ Aware Community in McKeesport, Pennsylvania. Located in an urban 10-acre, 12-block area. This is the first community-wide comprehensive research project. Although this community is still in a developmental stage, smart homes have been used for actual habitation. In addition, the community will have recreation facilities, retail and service providers on city streets, and public transportation (Wactlar et al., 2009). The purpose of this community is to

research individuals with decreases in instrumental activities of daily living (IADL). The infrastructure is designed to monitor activities inside and outside home.
In this community, Blueroof Technologies, a non-profit corporation, has developed the Smart Cottage for older adults. The cottage includes comprehensive home monitoring and Internet access designed for access, energy conservation, safety, convenience, and health maintenance. Eventually the community is planned to have over 100 residents occupying 40 dwellings, of which 60% will be older adults.
All homes in the community will be equipped with high speed Internet access and ZigBee monitoring which will be delivered by wireless transceivers. The transceivers will be integrated into a neighbourhood mesh that will deliver Internet access and energy control, media, and security monitoring to each resident in each home, which will be also connected through one private multifunctional network. Data collected from each home will be stored in a central server for data analyses. The 10 types of technologies used in each home are for: (1) energy management by wireless remote adjustment of thermostat, and automatic cutback on peak at night when unoccupied; (2) appliance (water and stove) and lighting control; (3) video monitoring via a front door camera modulated onto a TV channel; (4) video conferencing used for family communication, medical, and shopping purposes; (5) security for fire, smoke, and carbon monoxide alarms, window and door intrusion, and motion/temperature sensing; (6) health monitoring such as blood pressure, weight, medication management recording, and gross resident activity measures; (7) safety through activity monitors and alert buttons, and fall detectors; (8) wellness including diet, exercise, and preventive medicine; (9) Cyber Nurse allowing patient visits each day via the Internet, video conferencing to see and talk with patients, and recording patients' activities; and (10) Media Center to interface with the home's technology including an LCD TV/computer monitor.
Outside these homes, cameras will be installed throughout the community to monitor and record the activity of the community. Since this community level of intervention is a next generation project, only a brief description is made here. The purpose is for older adults with cognitive impairment to be able to participate in IADL such as shopping, crossing a road, and visiting someone by monitoring and supporting their activities. The main contributors of this project include researchers and developers from Carnegie Mellon University, Penn State University, and Blueroof Technologies, Inc.
Smart homes are becoming more inclusive of healthcare technologies and technologies to improve IADL, as well as even basic activities of daily living (BADL), in addition to the original purpose of added convenience. The previous examples of SH are newly built houses or residential complexes. However, the silver tsunami (baby boomers coming to age) will bring increasing demand for SH and healthcare technologies that can be added to existing structures because older adults wish to live in a familiar environment.

3.4 Cyber security issues

When a monitoring system is involved in a SH, the concern of security becomes a major issue. Due to the cyber-enabled mode of operation, information and network security are crucial issues for SH with healthcare technology. In addition to traditional physical home security requirements, SH adoption requires to solve novel cyber security vulnerabilities deriving from the network interconnection of different monitoring devices and SH terminals. Busnel (2008) has summarized the security requirements of a SH environment as follows:

Confidentiality: All communication between the SH terminals, monitoring devices and Emergency Response Centre (ERC) should maintain strict confidentiality. Patient's data available at SH terminals should be accessible to authorized requesters by enforcing standard authentication mechanisms.

Integrity: All the data being exchanged in SH environment should have guaranteed data integrity verification methods. Specially, data communication in SH environment is by and large wireless which is very susceptible to interception as it does not have the shielded protection of wired technology. Any data modified or altered during communication may cause inaccurate decision supplied for treatment planning and delivery.

Availability: The communication between the monitoring devices, SH terminal and ERC should be available on a continuous basis. The network could be proactively monitored for any disruption or anomaly in communication.

Therefore, non-compliance with the aforementioned requirements will leave the information in SH environment vulnerable to different cyber attacks. For instance, one threat to the availability requirement is a Denial of Service (DoS) attack which can lead to severe consequences. Proactive DoS attack mitigation techniques (Lee et al., 2008; Husain & Sridhar, 2009) should be adopted due to the time-critical nature of information exchanged in SH environment. Hash function based message authentication codes (Bellare, Canetti, & Krawczyk, 1996) can be used to maintain data integrity. To maintain data confidentiality, computationally efficient encryption solutions should be considered as most of the devices in SH environment are resource constrained (Oliveira et al., 2007; Sankaran, Husain, & Sridhar, 2009). These solutions, once customized for SH with healthcare technology, can ensure the necessary security required for optimal performance.

3.5 Retrofitted smart home and healthcare technology

Although individual cases of existing home retrofitting to smart homes may exist, there are few reports of systematic efforts made to convert to smart homes. Infrastructure wiring essential to SH technology is a major barrier to retrofitting older homes and apartments. For existing structures, among plug-and-play technologies, X10 devices have the longest history and been most extensively utilized.

The University at Buffalo, State University of New York, utilized X10 devices to retrofit 50 homes for older adults with chronic conditions living alone in their own home. X10 is a communications device that allows compatible products to connect to each other using the existing electrical writing in the home. Wireless motion detectors and hand-held remote controls utilize a radio frequency link. The detailed information for the installation and problems encountered for the process and their solutions can be found in the Aging and Technology Research website (www.agingresearch.buffalo.edu).

In the beginning, automatic, controlled lighting in each room was one of the most favoured smart home features. At the same time, lighting was the most troublesome feature because it was difficult to re-start after a power failure or when switches were inadvertently turned off. Door and window security, an automatic coffee maker, a security camera connected to lighting and alarm, and a medication reminder were other technologies utilized (Tomita et al., 2007). Its effectiveness has been tested, but this project did not incorporate healthcare

technology except for a personal emergency response system. The effectiveness of this system for their caregivers was measured (see the report later in this chapter).
The University at Buffalo also conducted a pilot study for e-health for patients with heart failure (Tomita et al., 2009). In this project, patients used the Internet and daily recorded their vital signs and health behaviours on a website that were monitored by health professionals. Changes in health status alerts were sent to the patient and health care provider. Participants in the intervention group had improved health outcomes and decreased emergency room visits. Smart home technologies and e-health each has unique contributions in the care of older adults, and there is important synergy when the technologies are combined. This will be discussed in a later section of this chapter.
Year 2009 was the first time the International Consumer Electronics Show featured a special section devoted to smart home or high-tech living for seniors. An important example other than motion sensors is a talking pill box that reminds users to take their medicines and which notifies caregivers of omissions (Zagier, 2009). The system utilizes ZigBee wireless technology to receive data from medical sensors. ZigBee consumes low power to transmit data over long periods of time making it suitable for thermometers, blood-pressure monitors, and pulse oximeters. Tele-conferencing between patient and doctor can occur for health status changes detected with the health monitors. Usefulness of these technologies may be self-evident, but there is little research supporting assumptions about effectiveness, and cost-effectiveness. Until there is more research, the role of the technologies and the optimal patient population for the technologies will remain unclearly defined.

4. Benefits of smart home and healthcare technologies for older adults

In this section, older adults' perception of SH and healthcare technologies, and the evidence in support of the technologies' effectiveness is discussed.

4.1 Perception of smart home and monitoring technologies

Studies on end users of SH are scarce. Since smart home users are still a small portion of the U.S. population, a survey (Mann at al., 2007) found that 98.2% of 673 respondents were nonusers of the technology. Calling it home automation systems, the study found that 66.6% of 661 respondents were not familiar with the system, 56.3% did not think it would be beneficial to them, and 59.3% were not interested in owning the system even after receiving an explanation. The major reasons for the technologies' perception as not beneficial were lack of perceived need, high cost, and unfamiliarity. Only 1.8% of the respondents used smart home technology. Among respondents with SH technology, 75% found it very important, and 71.4% were very satisfied. The results indicate that once older adults use SH technology, their perception of its importance and satisfaction will be high.
The researchers of the University of Missouri-Columbia have reported older adults' perceptions or attitudes toward the technologies. In the aforementioned assisted living place at the TigerPlace II, focus group sessions of 14 older adults (65+) were conducted to find their perceptions of the smart home and healthcare technologies before the installation (Demiris et al., 2008). Overall perceptions of the system applications were mostly useful, but most participants did not agree to the installation of video sensors in their own home because they felt it was unnecessary and an invasion of privacy. Specific responses for each feature were identified. The bed sensors were perceived overall as useful. The gait monitor

was perceived as very useful, as most participants expressed concerns about falling and being helpless or not detected in a timely fashion. The video sensor was perceived as beneficial in detecting health emergencies, such as falls. However, even if a video sensor is invisible to participants, they did not want to install it in their own homes. The researchers concluded older adults welcome emergency detection devices, but not home features for prevention through early detection of potential health problems. Participants who had previous experiences with health problems tended to perceive the benefits of these technologies.

Based on their comments, the researchers used monitoring sensors instead of cameras, and followed up with nine residents who lived in the SH. The technology used in the SH consisted of a set of wireless infrared proximity sensors to detect motions. They used low-cost X10 technologies coupled with specialized filtering and analysis. These sensors were installed in existing apartments of older adults who agreed to do so for the purpose of research. They reported the findings from the participatory evaluation of a SH application between Januray 2005 and August 2007 (Demiris et al., 2008). The researchers identified three phases in the process of adoption and acceptance of the technologies: 2-3 weeks for familiarization of technologies, 2-3 weeks for adjustment and curiosity, and full integration as the last phase. In the last phase, residents forgot the existence of the technologies, and the technology did not interfere with daily activities or cause privacy concerns.

In the 29th Annual International Conference of the IEEE, opinions of 30 leaders in the field of aging in the Northeastern United States and Wasthington D.C. were reported (Coughlin et al., 2007). The technologies were considered as having considerable potential to improve one's safety, but at considerable cost to independence. Constant monitoring as a way to ensure safety and security was regarded as a threat to dignity in one's own home. Unless individuals are extremely frail and the only other alternative may be nursing home placement, the idea of 24/7 monitoring was not acceptable. Therefore, smart technologies in the home may be a symbol of frailty rather than support for independence and health.

4.2 Effectiveness of smart homes

University at Buffalo, New York, using mostly low-cost X10 technologies, retrofitted 50 existing homes of older adults with chronic conditions who lived alone (Tomita et al., 2007). This study was the first randomized controlled trial (RTC) to evaluate the effectiveness of SH use in the United States. Participants selected the smart home technologies to suit their needs in the home. The project did not use continuous monitoring. All smart homes were equipped with window and door security, and an automatic lighting system operated through the participant's PC where ActiveHome (X10) software was installed. Many participants requested a chime module that was used for an alarm and a medication reminder. The intervention was for two years. At post-test, physical and cognitive functional status of SH users was maintained while those for the control group had significantly declined ($p<.05$). The difference was apparent in physical dysfunction, instrumental activities of daily living (IADL), mobility, and cognition. At the follow-up, the original 113 participants had reduced to 78. Among the treatment group, 80.4% were living in their own home compared with 65.7% for the control group.

This study also included participants' two follow-up (1 year and 2 years) perceptions of SH. Overall usefulness was positively perceived by 70.6% at the end of one year and 97.1% at the end of two years. Increased security and safety was the most frequent reason to recommend the technology to other people. This may be the reason for increased mobility. The additional positive impacts were knowledge gain and increased mental health (eg., increased mental stimulus and decreased levels of depression), and increased socialization via the Internet.

SH has limited implementation in the United States with consequent limited evidence of effectiveness. Furthermore, cost effectiveness studies and studies on appropriate technologies for specific populations or groups of older adults are needed.

4.3 Internet-based healthcare technology and its' perception and effects

While smart home technologies tend to focus on the prevention of falls and detection of illnesses based on older adults' body movements and activities, Internet-based healthcare technology primarily focuses on treating illnesses with prevention of complications as a secondary aim. This technology is variously termed as e-health, Internet-based self-monitoring system, Internet-based disease management, home health monitoring system, Internet-based telemedicine and telecare, etc.

In the United States, Internet-based disease management grew rapidly in the year 2000 in the healthcare industry. At that time, companies estimated significant cost savings. For example, traditional call-center programs cost $300 to $1,000 per patient per year, while a Web-based program was estimated at about $50 (Managed Care Week, 2000). The Internet continues to be regarded as an excellent compliment to an illness management program. Managed care companies initially pursued disease management for the high-volume chronic diseases diabetes, asthma, and heart disease. A model of disease management from Predictive Services, Inc. utilizes the following process: a virtual counselor guides the participant through an online baseline assessment, who then receives a personalized interactive session weekly. The session is data driven. The patient submits data that is sorted, filtered, and scored to generate personalized responses. This technology has been used by companies for weight management and fitness, pediatric and adult asthma, and women's and men's health.

In contrast to studies on SH effectiveness, there are many studies on efficacy or effectiveness of Internet-based healthcare technologies ranging from nonusers' perceptions of the technology to its effects on symptom management of specific illnesses. However, the vast majority of studies on the effectiveness of these technologies target younger adults.

Regarding perception of healthcare technology by older adults, survey results by Mann et al. (2007) found older adults perceived technology as useful for monitoring blood pressure (48.2% of respondents), 21.4% for blood glucose level, 14.4% for weight, 5.2% for heart rate, 2.8% cholesterol level, 2.4% each for mobility problems and falls, and 1.4% each for bladder function and pulmonary problems.

Research on the effectiveness of Internet-based health management shows its success in illness management. A randomized controlled trial was conducted to evaluate the immediate and long-term efficacy of a 12-month web-based intervention for improving diabetes outcomes (Bond, 2005). The study determined intervention effectiveness by measuring physical (glycosylated haemoglobin, weight, and lipid levels), behavioural (monitoring blood glucose levels, feet inspections, diet, and exercise frequency) and psychosocial factors (depression, quality of life, social support, and adjustment to diabetes). The average age of the participants (N=62) was 65.9 years old. All 31 participants in the treatment group were provided with a PC-based computer, a 17 inch monitor, a printer, and dial-up access to the Internet. Intervention participants could access online articles and websites on diabetes and other health related topics, received on-line advice, counselling, and encouragement from a nurse via e-mail, and could participate in weekly nurse-led chat/discussion and a peer support question and answer problem-solving forum. They received instruction regarding the development of personal action plans for diabetes management. These participants were requested to submit vital signs daily (blood sugar levels, medication administration, meal intake, weight and blood pressure). An Internet bulletin board offered the latest news in diabetes management. After five months, 10 of 15 participants decreased their average blood sugar readings by 15 points. All of them increased their exercise duration by 38 minutes on average, and they lost weight of 2.9 pounds on average, decreased in caloric intake by 336 calories, and lowered systolic and diastolic blood pressure by 10 and 5 points, respectively.

Similarly, Internet-based technology was used to promote self-management of patients with heart failure (Tomita et al., 2009). In this pilot study (N=40), all participants were 60 years of age or older. Patients in the treatment group were provided with a PC with Internet and training for computer use. The intervention was intended to empower patients with support from healthcare professionals using the Internet. Self-report of vital signs and targeted exercise on a secure web-site daily made them aware of the causal relationship between health behaviours and symptoms such as body weight and swelling. This RCT identified that after one year, knowledge of heart failure and related healthy behaviour and exercise frequency were increased, illness specific symptoms were improved, and blood pressure and quality of life were improved for the treatment group. Healthcare utilization was decreased for the treatment group. This study attributed the high adherence rate (85%) of the program to a simple web-based recording system, automatic alert for sudden weight gain or loss, patients' awareness that healthcare providers are watching their records, and healthcare professionals' personal feedback of the record.

On a very large scale, a one-year RCT was conducted at Stanford University to test efficacy of the use of Internet-based Chronic Disease Self-Management Program (CDSMP) to change health-related behaviours and improving the status of patients with chronic diseases such as heart disease, pulmonary diseases, or type 2 diabetes (Lorig et al., 2006). They tested whether an Internet-based method has the same effectiveness as the original small group method. Comparing 457 patients who used an Internet and e-mail method and a 501 patient usual care control, the study found significant improvements in health status in the treatment, compared with the control group. They concluded that the small-group CDSMP can be carried out by an Internet-based method and it is equally effective (Lorig et al., 2006).

A pilot study to compare two methods, Internet/personal digital assistant (PDA) versus face-to-face dyspnea self-management program, for 39 patients with chronic obstructive pulmonary disease (COPD) found that the effectiveness of both methods were similar (Nguyen et al., 2008). In this study, numerous technical challenges occurred with the PDA method and the project had to end earlier than the original plan; it was due to user unfriendliness of the website and PDA application difficulty. In academics as well as in industry, the rapid development of healthcare technologies is recognized; however for older adults, research on effectiveness and cost effectiveness is lacking.

5. Caregiver perspectives of SH with healthcare technologies

5.1 Caregivers for older adults in the United States

In the United States, 44 million individuals serve as informal and unpaid caregivers, and they provide 80% of the care for older adults, (National Alliance for Caregiving & AARP, 2004). Often informal caregivers are in a network including spouses, adult children, extended family, and/or friends. The increase in the number of older adults will hasten the need for informal caregivers due to the growth in the number of elderly individuals living with chronic and disabling conditions (Conway-Guistra, Crowley, & Gorin, 2002). Caregivers provide services ranging from bill payment, transportation, shopping, meal preparation, housework, to personal and complex medical care (Special Committee on Aging, 2001). Informal caregiving can positively impact the physical and mental health of older care recipients, but often at personal costs to the caregivers (Navaie-Waliser et al., 2002). Since caregiving can be stressful for caregivers, the needs of caregivers caring for older adults have been identified: to minimize caregiver burden and depression, and increase positive aspects of caregiving.

Many smart homes with monitoring systems include caregiver components. Healthcare technologies that are less costly and allow for remote caregiving along with safer independence for care recipients can be appealing to consumers. Cook (2007) notes that smart homes can be useful for improving the quality of life of older adults living at home alone in the community, while helping to alleviate the burden associated with caregiving. SH technology allows care recipients to be monitored within their home environment, while caregivers are able to receive health information that can be used to prevent or manage medical conditions. This concept has been implemented in many advanced SH with health monitoring systems.

For older adults living alone without cognitive impairment, personal emergency response systems (PERS) can notify a caregiver or agency that a fall has occurred. When the device is worn on the body, the wearer can press a button in the event of an emergency. Falls are the leading cause of home fatalities in older adults (Fuller, 2000), and a major concern among older adults driving a strong perception of need for these devices. Older adults'capacity to obtain help in times of distress can reduce worry and stress among caregivers and older adults alike, and may increase caregiver satisfaction. While much research focuses on the negative impacts of caregiving burden, the caregiving process can also include positive aspects. Interventions that promote positive attitudes and perceived satisfaction in the caregiving relationship are important in promoting health and family functioning, and are more likely to be adopted by consumers.

5.2 Smart home monitoring interventions and caregivers' perspectives

While there are a few studies on smart homes and their effects on older adults, only one research study exists regarding SH technology and the caregiver perspective in the United States. A study by Russ (2006) building upon the Smart Home Project by the Technology and Aging Research Project at the University at Buffalo, was conducted with 50 caregivers to older adults with chronic conditions who lived alone. These older adults had previously received computers with Internet capability and training for computer use, along with installation of the smart home (SH) X10 Active Home kit.

The intervention was comprised of installation and training of the X10 Powerhouse personal assistance security console system in the home, along with a wireless call pendant, together known as the PASS. These two devices utilized the telephone line to dial the caregiver for assistance and play back a stored message in the care recipient's voice. After pressing the call button, the console sounds an alarm to alert the user that the system is activated. The console dials and plays the message three times for the caregiver, and includes a statement asking the caregiver to press 0 on their touch-tone phone. If the listener presses 0, the alarm stops and he or she can listen for 75 seconds, at which time the care recipient can speak their message asking for help. If the phone has not been answered or an answering machine comes on, the console will dial the next of up to four stored telephone numbers. Seventy-five seconds after pressing 0, the alarm resumes and stops after four minutes unless the care recipient presses the STOP button on the call pendant, which will stop the alarm and reset the console. This stand-alone device was chosen because of its low cost and the relative ease of installation and use. The PASS had a onetime cost of $50.00 and no service or monitoring fees; more attainable for many people. Installation and initial training were provided. Ongoing technical support was provided, when necessary. A friend or family member was identified as the informal caregiver of interest for this study.

Caregivers were assessed for burden, depression, and caregiving satisfaction. The study compared the PASS users (treatment) and non-PASS users (control) who were further divided into e-mail users and non-users. The result showed that the PASS was effective in preventing an increase in caregiver depression and burden, but did not have any effect on caregiving satisfaction. E-mail use was found to be effective in preventing a significant increase in caregiver depression and burden in specific segments of the sample, but again had no effect on caregiving satisfaction. PASS and e-mail use combined was the most effective method to prevent increases in caregiver depression and burden over one year, while enhancing the relationship with the care recipient.

The primary concerns of the caregivers, for both the treatment and control groups (over 93% in both groups), were falls or injury as opposed to home security (less than 7%). One-hundred percent of the caregivers in the treatment group reported that they were pleased that their care recipients had the smart home, including the PASS, because of decreased worry about the care recipient's safety as a consequence of living alone.

PASS does not require the cost-prohibitive monthly monitoring fees seen with commercial personal emergency response systems. Cost is an important consideration, since over one-half of the care recipients had an annual income of less than $20,000. Older adults with disabilities and annual household incomes less than $20,000 were more likely to report inadequate assistance with activities of daily living and unmet needs. This puts these older adults at increased risk for adverse health outcomes and increased likelihood of institutionalization (Desai et al., 2001). The combined paid and unpaid costs to families for community care of frail older relatives ranges from $10,000 to $20,000 annually, depending upon their condition. These out-of-pocket costs may make formal care prohibitive (Allen, Foster, & Berg, 2001). Thus, the PASS is a feasible, low cost resource for these individuals. This technology is an option that can contribute to making the home a safe alternative to long-term care for many older adults.

A study from outside the United States was included here because of the paucity of data from the United States. Rialle et al. (2008) surveyed 270 families in France caring for persons with Alzheimer's disease and other dementia on their perceptions related to new smart home technologies to aid in dementia care at home. Two distinct groups were identified among family caregivers: one in favor of the use of technology, and those not in favors. Female caregivers were more willing to consider these technologies as an aid to caregiving. Although this result may be cultural, the study emphasizes the need to carefully consider the demographic characteristics of the caregiver in the development and targeting of healthcare technologies for use with older adults.

Increasingly, the perspective and the needs of caregivers are considered in the development of smart homes for older adults. This technology can potentially reduce caregiver depression and burden, and enhance the relationship with the care recipient. However, the benefits of smart homes may not be universal for all users, and are not fully understood.

6. The role of SH with healthcare technology and a conceptual framework

6.1 Technology that is needed

Physical and cognitive function is major determinants of an older adult's ability to live independently. The combination of the physiology of aging and chronic disease makes the maintenance of function a challenge with advancing age. The major illnesses affecting older adults' lives include arthritis, cardiovascular health, cancer, diabetes, epilepsy, obesity, oral health (Centers for Disease Control and Prevention, 2008). A healthy diet, smoking cessation, and exercise are important lifestyle elements that are part of disease management for many of the major chronic diseases. Promoting healthy behaviours should be a significant element of SH with healthcare technology.

Telehealth and e-health technologies specifically focus on monitoring vital signs for managing a chronic illness by healthcare professionals. Based on limited studies, older adults with chronic conditions accept monitoring for chronic conditions as opposed to lifestyle monitoring. In fact, the awareness that a health care professional is paying attention to the records is one of the reasons for high adherence to a health regimen. Even so, cost limits extensive implementation of the technology. What is lacking in SH with healthcare technology is inexpensive and secure technology that helps community-dwelling older adults with chronic conditions (1) manage their illnesses with monitoring by a healthcare professional and/or caregivers, and (2) maintain their physical and cognitive function without being monitored but connected to healthcare providers and caregivers. In addition, one of the most essential factors of the technology is that it should be incorporated into existing homes, so that older adults do not have to move to a new facility.

These technologies can be used for home therapy after surgery, but also need to be used after nurses' home visits end. The hospital stay time is becoming shorter due to changes in health insurance coverage policy. To prevent another hospitalization and emergency visits, and eventual nursing home placement, these technologies should be utilized daily for maintaining health, functional status, and quality of life or even improvement in older adults with various levels of conditions. It appears insufficient efforts have been made to provide these technologies for less frail older adults who do not require constant monitoring. The following section discusses the technology necessary to address needs of less frail older adults.

6.2 SH with healthcare technology

Although the optimal patient group has yet to be defined, one clear role for SH with healthcare technology is to enhance safety and promote physical and cognitive activities. The monitoring system should be developed with the user in mind. The system needs to be able to alert the user, as well as caregivers and first responders, when necessary.

Falls and safety prevention and detection: Sensors can detect when there is no movement or movement in an unusual location and issue a warning in an appropriate sequence. Or, a user may activate a personal alert.

Safety and security monitoring: It includes automatic control of water temperature, ambient temperature, stove and oven shut-off, lighting, window and door security, intruder alarm, and visitor identification. The frequency of a refrigerator door movement also can be recorded. Older adults may control timing and intervals of these devices through operation of a computer.

Health status monitoring: It can be also done by the SH such as under-the-bed sensors for detecting sleep patterns, a medication reminder, an appointment reminder, a toilet bowl sensor to check frequency of its use and health data such as the sugar level in urine. They may also include checking and recoding of vital signs such as blood pressure, glucose level, body temperature, pulse, weight, cholesterol, medication adherence, and interactions. A significant change in pattern or a significant delay in readings each serves as a trigger for an alert. Automatic transfer of vital signs has a purpose, but users benefit from knowing about and following the data. Audible and visual feedback from the measuring device can be very helpful when the user is monitoring his or her health status.

Nutrition/diet: One's caloric intake can be monitored and an audible and visual reminder alerts the person of a recommended consumption level.

Memory aids: They include reminder system announcing upcoming appointments or events and medication are useful for any older adult.

Physical activity level: This can be monitored without a video camera. This includes the amount of movements including exercise, ADL, IADL, and leisure. A wearable device informs about an exercise level and warns if there is evidence of excessive exercise. Video conference technology can promote a group exercise program and allow instructors to model the exercise. Group exercise and instruction may be particularly valuable during inclement weather.

The main concerns for these technologies are cyber security, privacy issues, costs, lack of human responders, and user friendliness combined with training.

Cyber security issues: SH with monitoring system is open to security vulnerabilities due to the network interconnection of different monitoring systems and SH terminals. Confidentiality, integrity, and availability have to be considered. Efficient encryption solution is essential.

Privacy issues: Strict control on the use and access to the monitoring data is an essential privacy standard. Older adults may become more involved if they control most of the monitoring data. The monitoring can give the older adult point-of-care feedback about their daily health status and prompt access to a health professional if that is warranted.

Costs: When the reporting goes directly to the user, the costs are much lower than when external entities participate in continuous monitoring.

Lack of human responders: The absence of a routine external monitor may leave users with the perception that there is no human invested in the process. Prompt response to alerts can allay some of those fears.
User friendliness: Older adults need simpler and less complicated interface and operation; training is another important element.

In summary, the role of SH with healthcare technology is to empower older adults to take care of their health and function themselves in their own homes. Most importantly, these technologies should promote healthy behaviours by raising awareness in older adults about the causal relationship between their daily activities and health and functional status. Figure 1 summarizes responsible parties for monitoring at the continuum of older adults' physical and cognitive function and their health status. Healthy older adults primarily doing self-monitoring is illustrated in the upper left in the diagram. The diagram illustrates a gradual decline in function moving clockwise. Ultimately, health monitoring may become the exclusive responsibility of professional caregivers for some older adults.

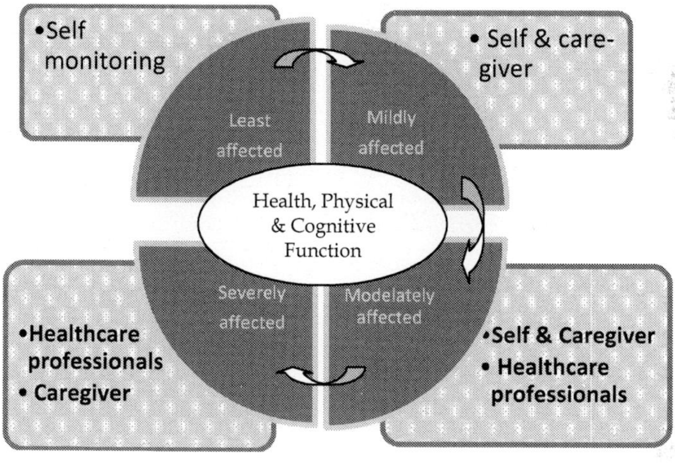

Fig. 1. Responsible parties for older adults' activity monitoring in relation to their health and physical and cognitive function

7. Model toward Optimal Management for Independence through Technological Adoption

A Model toward Optimal Management for Independence through Technological Adoption (M-TOMITA) had been proposed by the two authors in this chapter to identify the effectiveness of e-health for older adults with heart failure (Tomita et al., 2009). The model is an application of the original social support theory, to human support through technology utilization. It illustrates the role of e-health as a means to provide four types of support: informational, instrumental, emotional, and appraisal. The purpose of the support is to assist an individual to advance five stages of health behaviours in Transtheoretical model

(TTM): precontemplation, contemplation, preparation, action, and maintenance to manage their illness or health. The advantage of use of technology is to give users power to use services and information provided through technology at any time and any place. Also it gives users confidence that they can use advanced technology, can manage their health, and can connect to healthcare professionals instantly. This concept is based on mass media's Uses and Gratification theory.

These same concepts can be applied for SH with healthcare technologies. In this case, the purpose is not only to manage chronic illnesses, but also to maintain and improve functional status. SH should provide older adults feedback about health patterns and functional status which will lead to increased understanding of behaviour. This is informational support. SH should allow their review of records or input about activities and health. The involvement in this process is important so there is awareness of health/function behaviour, but also an understanding of cause and effect between behaviour and health/function. Instrumental support provides automatic alerts to the user, caregivers, healthcare providers and first responders. SH technologies provide appraisal support by evaluating one's behaviour, but if an automatic feedback is programmed, emotional support may also be provided by caregivers, and healthcare professionals provide emotional support. This process is continuous through SH with healthcare technologies.

The difference between the model's application to e-health and SH with health technology is that the latter may not be helpful to advance older adults stages of health behaviours from precontemplation, contemplation, to action. But the technology may be effective in advancing the stage from action to maintenance stage. Another advantage of this type of technology is that it is easy to make all readings and records available to caregivers. Even if an older adult is not motivated to self-manage/monitor behaviours, there is an opportunity for caregivers to assist with the monitoring. For willing caregivers, the technology may reduce adverse psychological effects of caring for older adults. The model is summarized in Figure 2.

8. Conclusion

Although the constant external monitoring (by others) feature of SH is useful for frail elders in assisted living facilities and nursing homes, it is not likely to be accepted by, and available for the majority of older adults who live in their own homes. Inexpensive SH technologies that can be used for the purpose of self-monitoring of safety, health, and functional status in existing homes are urgently required. The role of these technologies is to empower older adults for self-management for independent living and to minimize adverse effects on their caregivers' stress. The built-in warning system based on readings/reports collected by SH with healthcare technologies should send automatic alarms to self, caregivers, healthcare providers, and/or a fire department. The possibility of these SH with healthcare technologies is bountiful, but the research on its effectiveness and cost-effectiveness should be conducted concurrently, as they develop.

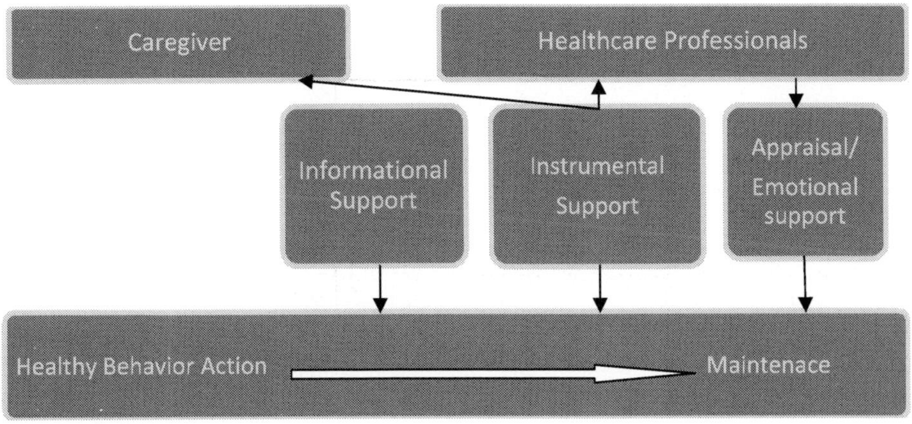

Fig. 2. Model toward Optimal Management for Independence through Technological Adoption applied for smart home with healthcare technologies

9. References

Administration on Aging, A profile of older Americans: 2008 (2009). Retrieved on September 5, 2009 from http://www.aoa.gov/AoARoot/Aging_Statistics/Profile/2008/14.aspx

Allen, S. M., Foster, A., & Berg, K. (2001). Receiving help at home: The interplay of human and technological assistance. *Journal of Gerontology*, Vol. 56B, (November, 2001), S374-S382, ISSN 1079-5014

ANI (2009) Health-monitoring system helps maintain older adults well-being, Thaindian News, January 7, 2009, retrieved on July 30, 2009 from http://www.thaindian.com/newsportal/health/health-monitoring-system-helps-maintain-older-adults-well-being_100139379.html

Bellare, M., Canetti, R., & Krawczyk, H. (1996). Meassage authentication using hash funcitons: The HMAC construction, *CryptoBytes*, LNCS 1109, Vo. 2 No. 1, 1-15

Bond, G. (2005) The challenges associated with implementing a web-based intervention for older adults: lessons learned. Retrieved on July 10, 2009 from http://faculty.cua.edu/tran/gsa-tag/GSA-TAG05/Bond-GSA05-WebIntervention.pdf

Busnel, P., El-Khoury, P., Giroux, S., & Keqin, L. (2008). Acheiving socio-technical confidentiality using securty pattern in smart homes. Proceeding of the Second Internationl Conference on Future Generation Communicaiton and Networking, Vo. 2, p.447, ISBN 9781424440085Hainan Island, China, December, 2008, IEEE Computer Socity, California, USA

Center for Future Health, (2005) Smart Medical Home research laboratory. University of Rochester. Retrieved on August, 20, 2009 from http://www.futurehealth.rochester.edu/smart;home/

Centers for Disease Control and Prevention (2008) Health information for older adults, Retrieved on July 10, 2009 from http://www.cdc.gov/aging/info.htm#2

Conway-Guista, F., Crowley, A., & Gorin, S. H. (2002). Crisis in caregiving: A call to action. *Health and Social Work*, Vol.27, No.4, (November, 2002) m307-311, ISSN 0360-7283

Cook, D. (2007) Providing for older adults using smart environment technologies, IEEE-USA Today's Engineer, May, 2007, retrieved on June 30, 2009 from http://www.todaysengineer.org/2007/May/smart_homes.asp

Coughlin, J., D'Ambrosio, L., Reimer, B., & Pratt, M. (2007). Older adults perceptions of smart home technologies: Implications for research, policy & market innovations in healthcare, Proceedings of the 29th Annual International Conference of the IEEE EMBS, August, 2007, 1810-1815, ISSN 1-4244-0788-5/07

Crawford, M. (2009). "Smart Homes" lay foundation for aging in place, FutureAge, July/August, 2009, 36-38. Retrieved on August 3, 2009 from http://cc.bingj.com/cache.aspx?q=smart+homes+%22lay+foundation+for+aging+in+place%22&d=76817406437344&mkt=en-US&setlang=en-US&w=418dc649,fef86ca6

Demiris, G., Oliver, D., Dickey, G., Skubic M., & Rantz, M. (2008). Finding from a participatory evaluation of a smart home application for older adults, *Technology and Health Care* Vol. 16, 111-118. ISSN 0928-7329

Demiris, G., Hensel, B., Skubic, M., & Rantz M. (2008). Senior residents' perceived need of and preferences for "smart home" sensor technologies, *International Journal of Technology Assessment in Health Care*, Vol. 24, No. 1. 2998, 120-124, ISSN 0266-4623

Desai, M. M., Lentzner, H. R., & Weeks, J. D. (2001). Unmet needs of personal assistance with activities of daily living among older adults. *The Gerontologist*, Vol. 41, No. 1, (February, 2001), 82-88, ISSN 0016-9013

Do, E.Y., Unlocking human potential through technological innovation- Interdisciplinary research at Georgia Institute of Technology. Retrieved on June 30, 2009 fromhttp://www.cc.gatech.edu/~ellendo/unlocking-GT_3.pdf

Foster, R. (2008). 2008 Medicare Annual Report, Financial Portion. U.S. Department of Health & Human Services, (April 01, 2008). Retrieved on September 5, 2009 from http://www.hhs.gov/asl/testify/2008/04/t20080401b.html

Fuller, G. (2000). Falls in the elderly, American Family Physician, April, 2000, Retrieved on July 20, 2009, from http://web.aafp.org/afp/20000401/2159.html ISSN 002-838X

Georgia Institute of Technology (2009) Aware Home Research Institute. Retrieved on July 27, 2009 from http://awarehome.imtc.gatech.edu

Helal, S., Schmalz, M., & Cook, D. (2008) Smart home-based health platform for behavioral monitoring and alteration of diabetes patients, *Journal of Diabetes Science and Technology*, Vol. 3, No. 1., 1-8, ISSN 1932-2968

Helal, S., Mann, W., El-Zabadani, H., King, J., Kaddoura, Y., & Jansen, E. (2005). The Gator Tech Smart House: A programmable pervasive space, *IEEE Computer*, Vol. 38, No. 3, (March, 2005) 64-74, ISSN 0018-9162.

Husain, M. I. & Sridhar, R. (2009). D-RNA: Towards a DDoS Resistant Network Architecture using Social Network Analysis. Proceeding of the ISCA 22nd International Conference on Computer Applications in Industry and Engineering (CAINE'09), November, 2009, San Francisco, California, USA, ISCA, North Carolina, USA

KUOW News (2008) Smart homes to help elderly remain independent longer, November, 13, 2008. Retrieved on June 30, 2009 from http://www.kuow.org/prgram.hhp?id=16299.

Lee, K., Kim, J., Kwon, K. H., Han, Y., & Kim, S. (2008). DDoS attack detection method using cluster analysis. *Expert Systtems with Applicaitons*, Vol. 34, No. 3, (April, 2008) 1659-1665, ISSN 0957-4174

Lorig, K.,Ritter, P., Laurent, D., & Plant, K. (2006). Internet-based chronic disease self-management: a randomized trial. *Medical Care*, Vol. 44, No. 11, 964-71, ISSN 0025-7079

Managed Care Week (2000). Internet-based disease management poised for take-off, Managed Care Week, July 10, 2000, retrieved on July 10, 2009 from http://www.mcareol.com/mcolfree/mcolfre1/artcl604.htm

Mann, W., Belchior, P., Tomita, M., & Kemp, B. (2007) Older adults' perception and use of PDAs, home automation system, and home health monitoring system, *Topics in Geriatric Rehabilitation*, January/March, 2007, Vol. 23, No. 1, 35-46, ISSN 0882-7524

Medical News Today. (2009), Seniors live at home longer with health-monitoring technology, MU researchers find. January 7, 2007, Retrieved on June 30, 2009 from http://www.medicalnewstoday.com/printerfiredly news.php?newsid=134639.

National Alliance for Caregiving & AARP (2004), *Caregiving in the U.S.* NAC and AARP, Retrieved on June 30, 2009 from http://www.caregiving.org/data/04finalreport.pdf

Navaie-Waliser, M., Feldman, P., Gould, D., Levine, C., Kuerbis, A., & Donelan, K. (2002). When the caregiver needs care: The plight of vulnerable caregivers, American Journal of Public Health, March 2002, Vol. 92, No. 3. 409-413. ISSN 1541-0048

Nguyen H., Conesky-Cuenco, D., Wolpin, S., Reinke, L., Benditt, J., Paul, S., & Carrieri-Kohlman, V. (2008). Randomzied controlled trial of an internet-based versus face-to-face dyspnea self-management program for patients with chronic obstructive pulmonary disease pilot study, Journal of Medical Internet Research, April/June 2008, Vol. 10, No. 2:e9 ISSN 1438-8871

Oliveira, L. B., Aranha, D. F., Morais, E., Daguano, F., Lopez, J., & Dahab, R. (2007). TinyTate: Computing the tate pairing in resource-constrained sensor nodes. Proceedings of the Sixth IEEE International Symposium on Network Computing and Applications (NCA 2007). pp. 318-323, IEEE, California, USA

Pinkus, A. (2009) Walking may foster friendships and achieve health. ABC News, (July 22, 2009). Retrieved on July, 22, 2009 from http://abcnews.go.com/GMA/VitalityProject/story?id=8140599&page=1

Rialle, V., Ollivet, C., Guigui, C., Herve, C. (2008). What do family caregivers of Alzheimer's disease patients desire in smart home technologies? Contrasted results of a side survey, *Method of Information in Medicine*, Vol. 47, No.1, 63-69, ISSN 0026-1270

Russ, L.S. (2006). The effect of home monitoring technology on reducing burden in caregivers of older adults with disabilities. Unpublished doctoral dissertation, State University of New York at Buffalo.

Sanakaran, S., Husain, M. I., & Sridhar, R. (2009). IDKEYMAN: An Identity-Based Key Management Scheme for Wireless Ad Hoc Body Area Network. Proceeding of the Annual Symposium on Information Assurance (ASIA'09), 23-28, June, 200, Albany, NY, USA, ISBN 1646-3692. ASIA, University at Albany, State University of New York, New York, USA

Special Committee on Aging. (2001). Long-Term Care Report, United States Congress, June 2002, Retrieved on June 30, 2009, from http://www.gpo.gov/congress/senate/senate22cp107.html

Stefanov, D., Bien, Z., & Bang, W. (2004) The smart house for older persons and persons with physical disabilities: Structure, technology arrangements, and perspectives. *IEEE Transactions on Neural Systems and Rehabilitation Engineering,* Vol. 12, No. 2, June 2004, 228-250, ISSN 0272-1716

Tomita, M., Mann, W., Stanton, K., Tomita, A, & Sundar, V. (2007) Use of currently available smart home technology by frail elders: process and outcomes, *Topics in Geriatric Rehabilitation*, Vol. 23, No. 1, 24-34, ISSN 0882-7524

Tomita, M., Tsai, B., Fisher, N., Kumar, N., Wilding, G., Stanton, K. & Naughton B. (2009). Effects of Multidisciplinary Internet-based program on management of heart failure, *Journal of Multidisciplinary Healthcare,* Vol. 2, January, 2009) 13-21. ISSN 1178-2390

Wactlar, H., Walters, R., Bertoty, J., & Hauptmann, A. (2009) The Aware Community, *International Journal of Smart Home,* Vol. 3, No. 1, January, 2009, 57-65. ISSN 1975-4094

Zagier, A., (2009) High-tech sensors help seniors live independently, Current World News, January 23, 2009, Retrieved on June 30, 2009 from http://www.current-news.org/2009/01/23/high-tech-sensors-help-seniors-live-independently/

9

Memory management in smart home gateway

Beizhong Chen
Rutgers, the State University of New Jersey
USA

Ibrahim Kamel
University of Sharjah
UAE

Ivan Marsic
Rutgers, the State University of New Jersey
USA

1. Introduction

With the trend of making house-hold appliances network-enabled and the widespread use of the broadband connections to homes, Internet services are increasingly finding their way to home applications and gradually changing the traditional lifestyles. As opposed to connecting traditional computing devices such as PCs, laptops, and PDAs, the Internet is increasingly used for home appliances such as television sets, refrigerators, air conditioners and washers (Ishihara et al. 2006; Ryu 2006; King et al. 2006), and other remote services, such as remote patient monitoring. Remote medical diagnosis, and remote configuration and control of home appliances are some of the most attractive applications. In the entertainment field, there are several interesting applications, such as downloading movies on demand and an Electronic Programming Guide (EPG). Electric power companies are also keeping an eye on home networking because it will allow them to provide value-added services, such as energy management, telemetry (remote measurement), and better power balance that reduces the likelihood of blackouts. Consumer electronics companies have started to design Internet-enabled products. Merloni Elettrodomestici, an Italy-based company announced their Internet washer Margherita2000 (Jansen et al. 2006), that can be connected to the Internet through which it can be configured, operated, or even diagnosed for malfunction (Ishihara 2006). LG presented the GR-D267DTU Internet refrigerator, which contains a server that controls the communication to three other appliances and has full Internet capabilities. Matsushita Electric showed during a recent Consumer Electronic exhibition an Internet-enabled microwave oven, which can download cooking recipes and heating instructions from the Internet. All of these effort lead towards the so-called "Internet of Things," which is forecast to grow vastly larger than the current Internet.

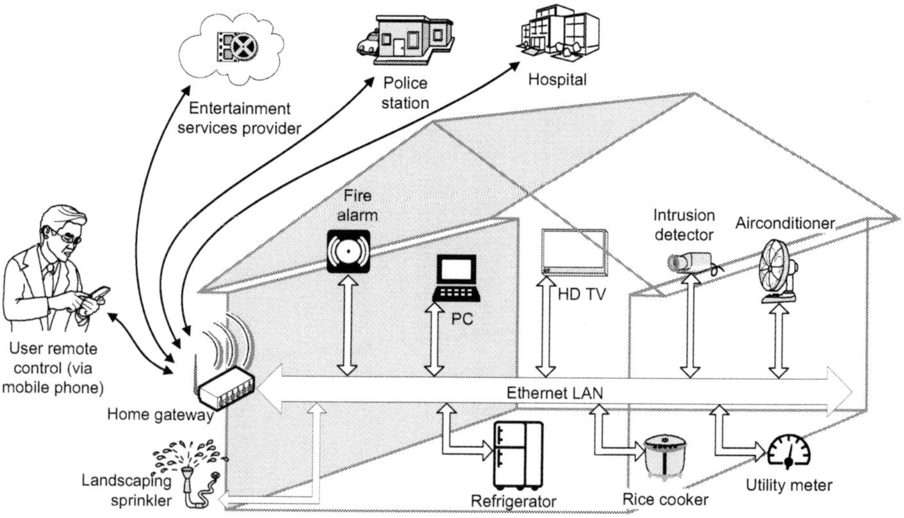

Fig. 1. A typical smart home network environment. See text for details.

Figure 1 illustrates a typical smart home network environment. Generally, there are several types of services in a digital smart home:

- **Basic services** (not shown in Figure 1):
  - Internet Access, Firewall, Personalized UI, Network Security, etc.

- **Information services and health care:**
  - Personalized e-Commerce services will automatically generate an order to a grocery shop when items in the refrigerator fall below a given threshold. This service can also support a more sophisticated shopping model that can bring improvements for the user, such as selecting a retail shop with the cheapest prices.
  - Remote Patient Monitoring will help reduce the cost of medical care. It mainly focuses on chronic diseases and monitors patient's physiological variables such as blood pressure, heart rate, blood sugar level, which will be recorded and transferred to a hospital for a timely professional consultation.
  - More information service models can be developed by service providers.

- **Communication:**
  - Unified Messaging services will allow users to have all the services that are expected in modern organizations, such as telephone call transfer, individual mail boxes, video sharing, etc. Voice over Internet Protocol (VoIP) can be part of this service, helping reduce the user's monthly telephone charges.
  - Other potential services include calendar, to-do list, e-mail, etc.

- **Entertainment:**
  - The Audio-on-Demand, Video-on-Demand, and Interactive TV bring to the home the newest popular music and movies. The users can select their favorites from a directory without leaving their couch. They can start, stop, or pause at will, just as they do for the discs or tapes at home.
  - Other potential services include updating electronic toys remotely, video games, virtual worlds, etc.

- **Control:**
  - Device Remote Diagnostics Service can test the appliances and system to ensure they are working properly. When a functional problem is detected, an alert can be sent. In the case when the problem cannot be solved remotely, the information will be sent to the service company and the service team can contact the user for proper parts. Device Integration Services allow the devices to communicate together and synchronize their work.
  - Energy Management can turn on the landscaping sprinkler when the power price is suitable, based on the weather. The user can use their mobile phone to send control signals to the air-conditioner at home to have it turned on before they arrive home.
  - More control services can be developed by appliances manufactures and service providers.

- **Home security:**
  - The home security systems, such as fire alarm system or intrusion detector, can be connected to a police station or security company to send alarm when unusual situation is detected for a quick response.

In this rapidly developing smart-home area, numerous companies compete to develop new technologies and products. Currently, there are several initiatives to define the specifications for network protocols and APIs suitable for home applications, such as Zigbee Alliance (2004), Sommers (2006), Jini by Sun Microsystems (2007), UPnP by Microsoft Corporation (2008), to name a few. It is expected that multiple home network protocols will coexist in the home and interoperate through the home gateway (Watanabe et al. 2007). Using a home gateway, previously isolated home devices such as digital camera, air conditioner, monitoring system, etc., can be integrated into a smart home system. The gateway also acts as a single point of connection between the home and outside world. Open Service Gateway initiative (OSGi) (The OSGi Service Platform Release 4 Core Specification version 4.2 2009; Binstock 2006) is a consortium of companies that are working to define common specifications for the home gateway. According to the OSGi model, the gateway can host services to control and operate home appliances. In the OSGi model, services are implemented in *software bundles* (or modules) that can be downloaded from the Internet and executed in the gateway (Maples & Kriends 2001). For example, HTTP service is implemented in one bundle, while security application could be implemented in another bundle. Bundles communicate and collaborate with each other through OSGi middleware and, therefore, some bundles may depend on other bundles. For example, a home security bundle uses an HTTP bundle to provide external connectivity (Lee et al. 2003).

To be deployed in customers' home, the price of the gateway is a main concern. Consumers might not be willing to pay for an extra box (home gateway). Also adding gateway functionality to an existing appliance, e.g., TV or set-top box (STB) would increase the appliance prices or shrink an already slim profit margin in this market. There is no consensus among consumer electronic industry on whether the gateway will be a separate box or it will be integrated in home appliances like digital TV or STB, or whether the gateway functionality will be centralized in one device or distributed among several appliances. However, the gateway will be, in general, limited in computational resources, especially main memory and CPU. The main memory in a home gateway will be used by various service bundles and home applications.

This chapter discusses the memory management in gateways and prioritizing the memory use to maximize the number of services running simultaneously in the home gateway. We propose new algorithms for efficient management of service bundles. Memory management has been studied extensively in the traditional operating systems field (Silberschatz & Peterson 1989). Due to different architecture, memory management for software bundles executed in home gateways differs from traditional memory management techniques in the following aspects:

- Traditional memory management techniques generally assume that memory regions used by different applications are independent of each other while some bundles may depend on other bundles in a gateway, as explained in Section 2.
- Due to the cost reason, many of the commercial gateways do not come with storage disks, which make the cost of stopping applications or services relatively high because restarting a service might require downloading the service bundle from the service provider through the Internet.
- Some home applications are real-time; therefore, removing a bundle from the memory may result in aborting the application or the service, while in traditional memory management model, removing a page from the memory costs only one disk I/O operation.

In a home gateway, generally, terminating a service might result in aborting one or more applications. However, in some applications it is possible to remove one service in the application and keep the application running. For example, audio-on-demand might still work without the equalizer service. However, if the application considers the terminated service critical to its operation, it might terminate all other services in the service tree as well. In this chapter, although the proposed model and algorithms work for the two cases mentioned above, we assume that terminating a node or a sub-tree from the service tree would terminate the whole application.

The main contributions of this work are:

- Identifying the difference between memory management in home gateway and traditional memory management problem in a general computing environment (addressed in operating systems literature).
- Introducing a model for the service replacement in home gateways using a directed dependency graph.

- Introducing SD (service dependency) Optimal algorithm which can work as a benchmark to compare different algorithms.
- Introducing SD Heuristic algorithm that performs significantly better than traditional memory management algorithms and close to the optimal solution based on Knapsack problem.

This chapter is organized as follows. The next section describes prior work, and introduces OSGi and the corresponding memory model. Section 3 presents an application scenario with counter example and a formal definition of the memory management problem. In Section 4 we first present the simple algorithms that are based on traditional methods and then propose our SD Heuristic algorithm and an analytical solution (SD Optimal) based on Knapsack problem. Evaluation results and discussion are presented in Section 5. Finally, conclusions and outlook are summarized in Section 6.

2. Prior work

Memory management has been discussed extensively in the operating systems literature. For the sake of comparison, we adopted two well-know memory management techniques, namely, best-fit, worst-fit and compared them with our proposed protocols in Section 4. Due to the different architectures, one of the main differences between memory management for smart home applications and general computer applications is that the former takes into account the dependencies among different services or bundles, as explained later. To our best knowledge, there is no study related to the memory management in the context of smart home applications. Vidal et al. (2006) addressed QoS in home gateway. They proposed a flexible architecture for managing bandwidth inside the home. However, they have not addressed memory management in home gateways. Aliet al. (2005) proposed architecture based on OSGi for wireless sensor networks, where data is processed in distributed fashion. They showed how to execute simple database queries like selection and join in a distributed fashion. Bottaro et al. (2007) addressed protocol heterogeneity and interface fragmentation when connecting several devices to OSGi-based gateway at home. The paper describes different scenarios and challenges for providing pervasive services in home applications.

2.1 OSGi framework introduction

Due to the different characteristics from traditional computer architecture, Ericsson, IBM, Motorola, Sun Microsystems found OSGi Alliance, which is an open standards organization in March 1999. Developed by this alliance, OSGi is a Java-based framework and (Helal et al. 2005; Lee et al. 2003) and wireless networks (Helal et al. 2005). The OSGi framework is completely based on Java technology. In fact, the specification itself is just a collection of standardized Java APIs plus manifest data. The use of Java technology has several important advantages. First, Java runtimes are available on almost all OS platforms, allowing the OSGi framework and services to be deployed to a large variety of devices across many different manufacturers. Java also offers superb support for secure mobile code provisioning, which allow developers to package and digitally sign a Java applications and send them over the network for remote execution. If the execution host cannot verify the digital signature or determines that the application does not have sufficient permission, it could reject the application or put it in a sandbox with limited access to local resources. Furthermore, Java has an extensive set of network libraries. It supports not only HTTP and TCP/IP networking, but also advanced peer-to-peer protocols such as Jini, JXTA and BlueTooth. Services are implemented as plug-ins modules called bundles. (We will use the terms "bundle" and "service" interchangeably in the rest of this chapter). These bundles can be downloaded from the application service providers through the Internet when they are requested. Examples for services that are used for application development are Java development tools, J2EE monitor, crypto services, bundles that provide access to various relational database management systems (e.g., DB2, Oracle, etc.), HTML creation, SQL, Apache, Internet browser, XML plug-ins, communication with Windows CE, etc. Other system administration bundles like core boot, web application engine, event handling, OSGi monitor, file system services, etc. Bundles for various Internet and network protocols, like, HTTP service, Web services, SMS, TCP/IP, Bluetooth, X10, Jini, UPnP, , etc. There are many bundles that are already implemented by OSGi partners (Binstock 2006).

2.2 Service dependency graph

Figure 2 shows the software architecture of a gateway according to the OSGi model. Some of the basic bundles, which implement essential services, are already loaded in the gateway framework. The framework handles the basic bundle management functionality, e.g., install, uninstall, start, stop, communication, etc. Other service bundles, developed by the third party like device manufacturers and services providers, can be downloaded in real-time by the OSGi framework as needed.

Our proposed algorithm is implemented as a part of the framework to provide memory management. The gateway can download the corresponding bundles (that correspond to specific services) when it becomes necessary. In order to share its services with others, a bundle registers any number of services to the framework.

A bundle may import services provided by other bundles and therefore, its running may depends on other bundles. For example, a file downloading bundle needs services provided by an UDP service bundle or TCP service bundle, therefore, it is dependent on UDP service bundle or TCP service bundle. To model the relationship among services, we use a dependency graph. Formally, given a set of service instances $S = \{ s_1, \ldots, s_n \}$ which currently reside in main memory, let $G (S,E)$ be a directed acyclic graph with vertex set S and edge set E, describing the dependence among these instances. There is a directed edge from s_{ij}, $(s_i,s_j) \in E$ if and only if s_i depends on s_j. Since it is natural to assume that each application instantiates its own copy of a given service, the dependency graph will consist of a forest of rooted trees, where each tree represents the service instances instantiated by a given application as shown in Figure 3.

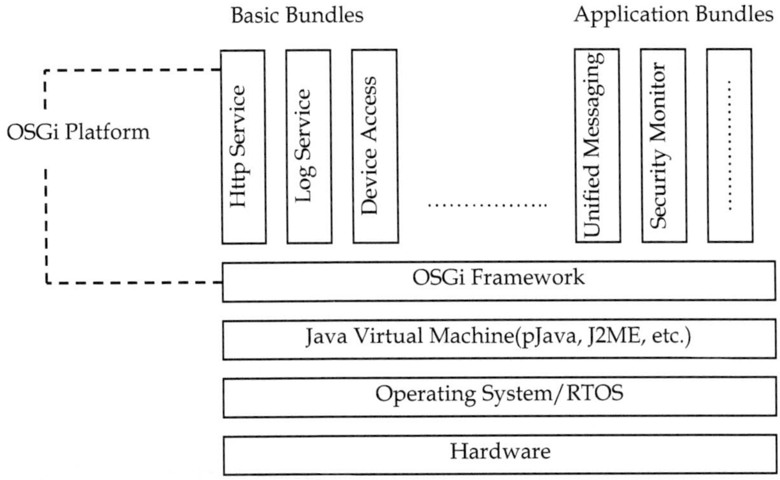

Fig. 2. Software architecture of a gateway in the OSGi model.

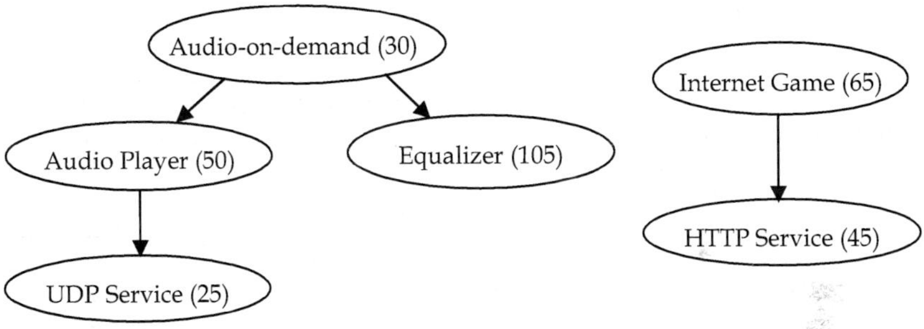

Fig. 3. Dependence graph for two currently running applications (numbers indicate memory requirements).

3. Problem definition

The gateway might need to free memory space to accommodate new services that are triggered by connecting a new device to the network or upon explicit local or remote requests. Although the amount of memory required to execute a service might change with time, the application service provider (or the author who provides the bundle) can give approximate statistical estimates of the amount of memory required to execute the services such as average, median, or maximum. Moreover, extra memory space might be requested by any one of the service instances (inside the home gateway) to continue its service. If such memory is not available, the gateway has to pick a victim service instance (or instances) to terminate to allow the new application to start. Note that because many of the smart home applications are real-time in nature, thus, the gateway tends to terminate the victim service rather than suspending it. Ideally, the gateway memory management algorithm needs to meet the following desirable properties:

- The total amount of reclaimed memory is enough to fulfill the requested memory.
- The number of victim service instances should be minimal.
- Since the algorithm will be executed in real-time, it should be fast and does not require much memory itself.

3.1 Application scenario

The problem addressed in this chapter can be better described by the following motivating example. Suppose that there are two applications that are running on the gateway. The first application uses audio-on-demand service that depends on the audio player service, which in turn, depends on the UDP service. The service dependency graph for the audio-on-demand is shown in Figure 3. The second application is an Internet game, which consists of two services; a game service that depends on HTTP service. Now we would like to start one more application, for example, home security. Let us assume that there is no more free memory in the gateway and the total memory required by the home security application is 100 (memory units).

Apparently, home security application is more important than Audio-on-demand application and Internet-game. Thus, it is reasonable to kick out at least one of these services to start the home security application. As shown in Figure 3, the equalizer service uses 105 memory units. A wise decision would be to kick the equalizer service that belongs to the audio-on-demand application because it results in killing less number of service instances and fulfills the memory demand.

The challenge is to select those services to kick out from memory such that the number of services and applications affected is minimal and that the total memory reclaimed equal or greater than the memory requested. In Section 4, we propose new algorithms for service replacement for memory management in gateways.

3.2 Formal description of the problem

More formally, our problem can be described as follows. Let $S = \{ s_1, \ldots, s_n \}$ be the set of service instances currently resident in gateway memory (Table 1). Service instance s_i occupies $M(s_i)$ storage. Let $G(S,E)$ be the forest of trees describing the dependency among the set of instances S. For a vertex v in G, let us denote by $T(v)$ the set of vertices of the sub-tree of G rooted at v (including v itself), and for a subset of vertices $V \subseteq S$, let $T(V) := U_{v \in V} T(v)$.

Given that a new service instance s, with memory requirement $M(s)$ has to be created, it might be required to remove some currently existing instances in order to free room for the new instance. Assume that the extra required memory for this operation is M_t units, that is $M_t = M(s) - M_f$, where M_f is the current amount of available memory. Here we assume that, when a service instance is terminated, all instances depending on it will be terminated and removed as well. Our goal is to reduce the number of removed (stopped) services. More precisely, it is desired to find a subset $V \subseteq S$ of minimal number of service instances, whose ejection, together with all its dependents, will make available a total memory of at least M_t units. Letting $M(S') := \Sigma_{s \in S'} M(s)$ for any $S' \subseteq S$, our problem can be formulated as finding

$$\min \{ |T(V)| : V \subseteq S, M(T(V)) \geq M_t \} \quad (1)$$

This last problem is closely related to the well-known Knapsack problem, which is NP-hard in general (Garey & Johnson 1979). However, the Knapsack problem admits a pseudo polynomial algorithm which runs in $O(n^2)$ (see, for example, Jain & Vazirani 2001). This solution is discussed in detail in Section 4.3.

Term	Description
Service	A self-contained component that performs certain functionality
Bundle	The functional and deployment unit for shipping services
Service instance	The execution thread of a bundle in the framework
Applications	Consists of one or more services

Table 1. Description of the terminology used in the chapter.

4. Service replacement algorithms

In this section we present several algorithms for solving service replacement problem in home gateway. The first three algorithms are direct adaptation of the well-known first-fit, best-fit and worst-fit algorithms which select the service(s) to be replaced based on the amount of memory that might become available. The other two algorithms take into account not only the memory size but also the service dependencies. SD (Size-Dependency) heuristic is a simple heuristic that runs in $O(nh)$ time and requires linear space, where h is the height of the forest. Finally SD Optimal algorithm computes an optimal solution in $O(n^2)$ time and $O(nh)$ space.

As explained in the previous two sections each application is modeled as a tree of service nodes that are used by this application. The algorithm showed in Figure 4 selects a victim node (root) X. Note that X can be either an application (a root of a tree) or a sub-tree that belongs to an application. If X is the root node then the gateway will stop the corresponding application. But if X was a sub-tree under an application, then deleting X might stop some functionalities of the application without terminating the whole application. Our proposed model and solution work equally for both cases.

In some special cases, the application might still run with reduced functionality as a result of stopping the sub-tree rooted at X. However, it is highly possible that the deletion of the sub-tree X seriously affect the execution of the application and consequently the whole application stops. Even though our algorithms work for both cases, in our experiments we count on the general case that deleting any sub-tree stops the whole application. Thus, without loss of generality, the following discussions and algorithms check only root nodes (application node) and not a sub-tree. In general service replacement algorithm is shown in

Figure 4. The differences between the different techniques are materialized in the way the algorithm selects the next victim for deletion in Step 2.1 in Figure 4. It is easy to see that the above heuristic can be implemented in $O(nh)$ time and $O(n)$ space. In the next two sections we discuss several alternatives for picking the victim bundle.

Generic (G, S, M_t)
Input: The current set of service instances S, the dependence forest G, and the memory requirement M_t.
Output: A new dependent forest G, describing the dependency among the bundles remaining after deleting a set of bundles whose total memory is at least M_t.

1. For each node $s \in S$, compute the accumulative size and memory:
   $c(s) \leftarrow |T(s)|$ and $m(s) \leftarrow M(T(s))$ using breadth first search on the dependency forest G
2. While $M_t > 0$
   2.1 Pick a victim node s, according to the selection strategy to be described later.
   2.2 Delete s and all its dependents
   2.3 For every node u on the path from s to the root of the tree containing s, set
   $c(u) \leftarrow c(u) - c(s)$ and $m(u) \leftarrow m(u) - m(s)$
   2.4 Update $M_t \leftarrow M_t - m(s)$

Fig. 4. General service replacement algorithm.

4.1 The simple algorithm

This algorithm is similar to the traditional algorithms used in operating system literature for memory management in general purpose computers. These traditional techniques make selection based on the amount of memory used and ignore the dependencies. We modify these techniques to take into consideration the total *accumulative* memory of each service (bundle) resulting from stopping one service. We consider the following three algorithms:

First Fit: choose the first service s in the list $S=\{s_1, \ldots, s_n\}$ such that total memory $M(T(s))$ occupied by its sub-tree is *at least* the requested amount M_t:

$$k \leftarrow \min\{1 \le j \le n \mid M(T(s_j)) \ge M_t\}; \qquad s \leftarrow s_k$$

If no such node exists ($k=\infty$), pick the node with largest $M(T(s))$:

$$s \leftarrow \text{argmax}\{M(T(s)) : s \in S\}$$

where argmax{ ... } denotes a maximum of a given function.

Best Fit: choose the service $s \in S$ with the smallest total memory that is $\ge M_t$:

$$s \leftarrow \text{argmax}\{M(T(s)) : s \in S, M(T(s)) \ge M_t\}$$

where argmax{ ... } denotes a maximum of a given function.
If no such node exists ($k=\infty$), pick the node with largest $M(T(s)$.

Worst Fit: choose the service $s \in S$ with the largest total memory:

$$s \leftarrow \text{argmax}\{\, M(T(s)) : s \in S\}$$

If no such node exists ($k=\infty$), pick the node with largest $M(T(s))$.

4.2 SD heuristic

Different from the simple algorithms discussed above, our proposed heuristic greedily tries to pick, as a victim for deletion, the service instance whose removal will free the minimum amount of memory larger than M_t and, at the same time, it has the smallest number of dependents. Towards this end, the heuristic will pick for deletion the service instance s which maximizes the ratio of the total memory to the number of dependents:

$$s \leftarrow \text{argmax}\{\, M(T(s))/\,|\,T(s)\,|\; : s \in S\}$$

This selection tends to decrease the number of deleted instances. Looking back at the example in Figure 3, we can see that the ratios $M(s^*)/\,|\,s^*\,|$ for the different service instances are as follows: Audio-on-demand (52.5 = (30 + 50 + 25 + 105)/4), Audio-player (37.5=(50+25)/2), UDP Service (25), Equalizer (105), Internet game (55=(65+45)/2), and HTTP Service (45). Thus, the service instance with maximum ratio is the Equalizer whose removal will give enough memory to start the new service (requiring 100 memory units). Should we have used the First Fit strategy, on the other hand, we might have selected to remove the Audio-on-demand instance, which results in removing four instances instead of only one. Note also that, for this particular example, the Best Fit algorithm would also remove the same instance (the Equalizer) selected by the SD Heuristic.

4.3 SD optimal

It is well known that the Knapsack problem admits a pseudo polynomial algorithm. In this section, we extend this solution to problem 1 using dynamic programming (Johnson & Niemi 1983). Specifically, let $S = S_n = \{s_1,\ldots,s_n\}$ be the current set of service instances listed in post-order traversal (that is, we recursively traverse the children from left to right then we traverse the root). We shall consider incrementally the sets $S1 = \{s_1\}$, $S2 = \{s_1, s_2\}$, $S3 = \{s_1, s_2, s_3\}$, ... , computing for each set the maximum amount of memory that can be achieved by deleting a subset of nodes. In order to compute these maxima, we will need to compute for each node s_i, the largest index $k \in \{1,\ldots, i-1\}$ such that s_k is *not a descendant* of s_i. Let $L(s_i)$ denotes such an index. The following procedure gives the post-order traversal of a given forest and computes the required indices $L(s_i)$ for each $i=1,\ldots,n$.

If the connected components of the forest G are $C1$, $C2$, ..., Cr, then in order to compute the post-order traversal for G, the above procedure is called r times.

Traverse (v, G, k)
Input: a sub-tree of the dependence forest G rooted at v and an integer k.
Output: the post-order traversal { s_k, s_{k+1},..., $s_{|T(v)|+k+1}$ }of T(v), and the set of indices {L(s) | $s \in T(v)$}

1. If |T(v)|=0 // *tree is empty*
   return
2. If |T(v)|=1 // *v is a leaf node*
   L(v) ← k
3. else for each child u of v:
   Traverse(u,G,k);
   L(v) ← L(leftmost(v))
4. k←k+1
5. s_k ← v

In what follows, nodes u, $v \in V(G)$, are considered to be incomparable if neither is a descendant of the other, i.e., $v \notin T(u)$ and $u \notin T(v)$. Note that n is a trivial upper bound on the total number of instances (or weight) that can be achieved by any solutions.

Traverse-Forest (*G*)
Input: the dependence forest G
Output: the post-order traversal $\{s_1,...,s_n\}$ of G, and the set of indices $\{L(s) \mid s \in V(G)\}$

1. Find the connected components *C1, C2, ..., Cr* of *G*
2. k← 0
3. For i =1 to r
   Traverse (root (C_i), G, k)

For each $i \in \{1,...,n\}$ and each $w \in \{1,...,n\}$, let $S_{i,w}$ denote a subset of incomparable elements of $S_i = \{s_1,..., s_i\}$, whose total weight is exactly w, and whose total memory is maximized. Let $A(i,w)=M(T(S_i,w))$ if the set $S_{i,w}$ exists, and $A(i,w) = -\infty$ otherwise.

$$|T(S)|=w,\ A(i,w) = \begin{cases} -\infty \text{ if the set } S_{i,w} \text{ does not exist} \\ 0, \text{ if } i=0 \text{ or } w=0 \end{cases}$$

Clearly $A(1,w)$ is known for every $w \in \{1,...,n\}$. The other values of $A(i,w)$ can be computed incrementally using the following recurrence:

$$A(i+1, w) = \max\{A(i,w), M(s_{i+1}) + A(L(s_{i+1}), w - |T(s_{i+1})|)\} \quad (2)$$

if $|T(s_{i+1})| \le w$ and $A(i+1,w)=A(i,w)$ otherwise.

Proof of Eq. 2: Let $S' \subseteq S_{i+1}$ be a subset of incomparable elements that achieves $A(i+1, w)=\max\{M(T(S)) \mid S\ S_{i+1}, |T(S)|=w\}$. There are two possible cases:

Case 1: $s_{i+1} \notin S'$. Then $S' \subseteq S_i$ achieves $A(i,w) = \max\{M(T(S)) \mid S \subseteq S_i, |T(S)| = w\}$.
Case 2: $s_{i+1} \in S'$. Let $S'' = S' \setminus \{s_{i+1}\}$. Since the elements of S′ are incomparable and the dependence graph is a forest, we have $T(S') \cap T(s_{i+1}) = \emptyset$, and therefore,

$$|T(S'')| = |T(S')| - |T(S_{i+1})| \text{ and } M(T(S'')) = M(T(S')) - M(T(s_{i+1})).$$

By the definition of $L(s_{i+1})$, we know that for $L(s_{i+1})+1 \leq j \leq i$, s_j is a descendant of s_{i+1}, i.e., $T(s_j) \cap T(s_{i+1}) \neq \emptyset$, implying that S'' must be a subset of S_k, where $k=L(s_{i+1})$. Thus $S'' \subseteq S_k$ is a subset that achieves $A(i,L(s_{i+1}))=\max\{M(T(S)) \mid S \subseteq S_k, |T(S)| = w - |T(s_{i+1})|\}$, which when combined with s_{i+1} gives $M(T(S'))=M(T(S''))+M(T(s_{i+1}))=A(i,L(s_{i+1}))+M(T(s_{i+1}))$.

Equation 2 then follows by taking the maximum achievable memory over cases 1 and 2. □

Now we state the optimal algorithm.

Optimal (G, S, M_t) The current set of service instances S, the dependence forest G, and the memory requirement M_t.

Output: A new dependence forest G, describing the dependence among the bundles remaining after deleting a set of bundles whose total memory is at least M_t.

1. For each node $s \in S$, compute the accumulative size and memory:
   $c(s) \leftarrow |T(s)|$ and $m(s) \leftarrow M(T(s))$
2. Call Traverse-forest(G) to get the post-order traversal $\{s_1, \ldots, s_n\}$ of G, and the set of indices $\{L(s) \mid s \in V(G)\}$.
3. Initialize:
   $A(i,0)=0$ for all $i=1,\ldots,n$,
   $A(0,w)=0$ for all $w=1,\ldots,n$,
   $A(1,1)=m(s_1)$, and $A(1,w)=-\infty$ for all $w=2,\ldots,n$.
   // Build a dynamic programming table
4. For i=1 to n
5. For w=1 to n
   if $c(s_{i+1}) \leq w$
     if $A(i,w) \geq m(s_{i+1}) + A(L(s_{i+1}), w - c(s_{i+1}))$
       $A(i+1,w) \leftarrow A(i,w)$, $B(i+1,w) \leftarrow 0$
     else
       $A(i+1,w) \leftarrow m(s_{i+1})+A(L(s_{i+1}),w - c(s_{i+1}))$, $B(i+1,w) \leftarrow 0$
   else
     $A(i+1,w) \leftarrow A(i,w)$, $B(i+1,w) \leftarrow 0$.
   // now compute optimal solution
6. $S \leftarrow \emptyset$; $i \leftarrow n$; $k \leftarrow \min\{w \in [n]: A(i,w) \geq M_t\}$.
7. while $i > 0$
   if B $(i,k) = 1$
     $S \leftarrow S \cup \{s_i\}$; $i \leftarrow L(s_i)$; $k \leftarrow k - c(s_i)$.
   else
     $i \leftarrow i-1$.
8. For each $s \in S$, delete $T(s)$.

Thus we get an $O(n^2)$ time, $O(nh)$ space algorithm for solving problem 1.

5. Performance evaluations

We carried extensive studies to evaluate the proposed algorithms. First, we compared the performance of the different algorithms in terms of the number of removed services to verify our new proposed algorithms. And then evaluate the algorithm execution time to show that the SD heuristic is practical in a home gateway. We considered different scenarios e.g., different distributions of bundle (or service) sizes, different number of existing bundles, etc. First we describe how the experimental data is generated, and then we present our results.

5.1 Experiment setup

Initially, services are generated with random sizes and loaded into the gateway memory, until the memory becomes almost full. Each service can depend on a number of randomly selected services with probability varying from 0 to 1. Service sizes are selected randomly in the range from 100 Kb to 50 Mb according to different probability distributions: uniform distribution in the given range, exponential distribution with a mean 5M, and a normal distribution with a mean of 5M.

Because home gateways are new, it was difficult to find real data (traces) of the service arrival. In our experiments, we used statistical service arrival model. We used both uniform distribution and exponential distributions for new service arrival to the home gateway. We conducted experiments to compare the performance of the following algorithms:

- Traditional algorithms: Best-fit and Worst-fit
- SD heuristic
- SD Optimal algorithm

A new service, with memory requirement varying uniformly 100K–50M, is created. We find out which services (bundles) should be kicked out to make enough room for the incoming bundle. Two performance measures were considered:

1. The number of services need to be stopped (or kicked out) to free enough space for the new service
2. The cost of each algorithm, in terms of execution time, required to determine the victim services (bundles).

Each performance measure was averaged over 1,000 experiments.

5.2 Experimental results

In our first experiment, we fixed the number of existing bundles in the home gateway and then compared how the different algorithms behave in terms of the number of kicked out services, as the size of the new coming service (s_{new}) is increased from 100K to 50M. In all our experiments, we assumed uniform and exponential service arrival. However, service arrival distribution does not affect the number of victim services. In Figures 5, 6 and 8, service arrival is assumed to be uniform. Exponential distribution gives similar results and thus not shown. Figure 5 shows our results when the number of services currently running in the gateway=100. Just as we have expected, it can be seen from Figure 5, the SD heuristic and the SD Optimal perform much better than the traditional techniques. This result verifies that our proposed algorithms perform much better than the traditional techniques, after taking the dependency between different bundles into account. We also note that the SD heuristic performs very close to the SD Optimal for various size of the new service s_{new}.

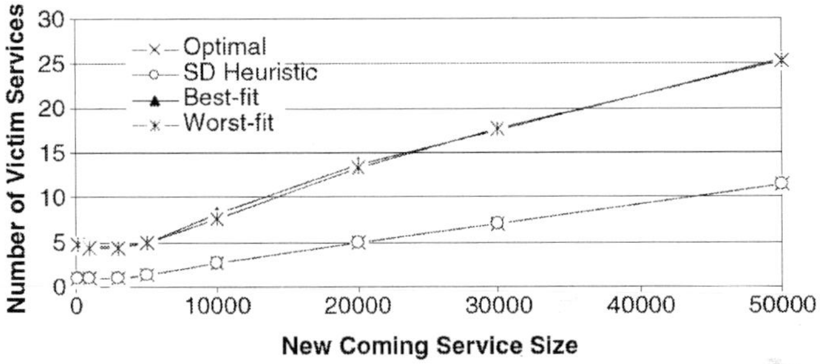

Fig. 5. Performance of the different algorithms as function of S_{new} for uniform distribution.

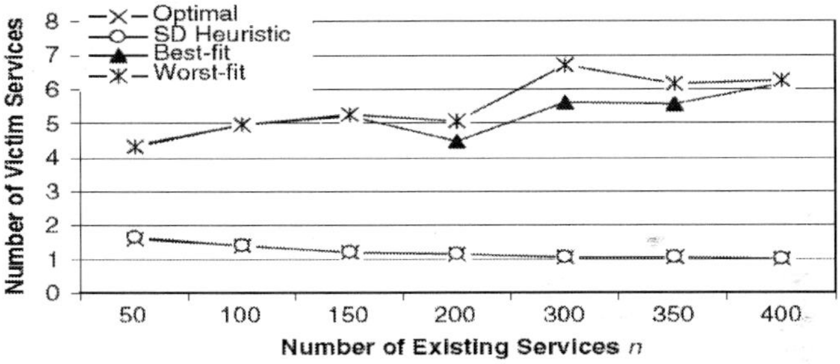

Fig. 6. Performance of the different algorithms as function of n for uniform distribution.

In the second experiment, we compare the performance of the different algorithms as the number of existing bundles n is increased. The result is shown in Figure 6. As we can see from the result, the performance of SD optimal and SD heuristic remain almost invariant under the change of number of bundles. The performance of the traditional techniques, on the other hand, degrades as the number of services running in the gateway increases. This can be explained as follows. With a large number of existing bundles, the chances that the memory requirement will be fulfilled with a few number of bundles from the lower levels (i.e., having a few levels of descendants) is higher. Since SD heuristics and SD optimal take dependencies into consideration, the likelihood to find better solution increases with the increasing of the number of existing services. Their performance will improve with the increase in chances of finding bundles which have less dependent bundles, and therefore, fewer services are terminated. On the other hand, the traditional techniques do not consider the dependencies between different services in the OSGi platform and provide no optimization, and therefore, might have to delete a few bundles from the top levels, resulting in a much higher number of kicked out bundles.

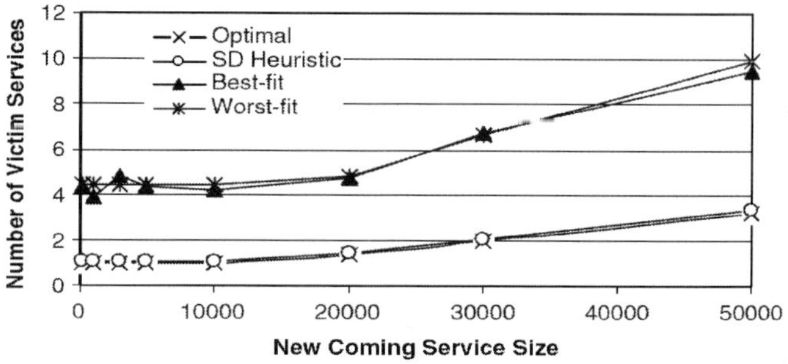

Fig. 7. Performance of the different algorithms as function of s_{new} for exponential distribution.

In the next experiment, we examined the effect of using a non-uniform distribution on the performance of the algorithms. We used an exponential distribution with mean 5M for the size of the existing bundles. Figure 7 presents our results for this experiment. Clearly, the number of kicked out bundles has decreased relative to the uniform case, since in this case it is easier to satisfy the memory requirement with a smaller number of bundles. However, we notice that the relative performance of the different algorithms remains invariant.

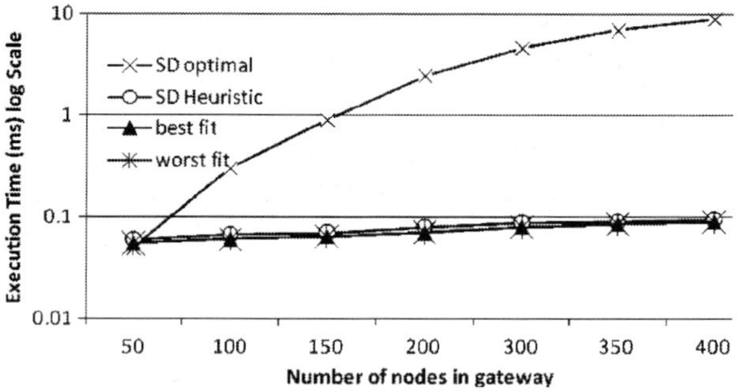

Fig. 8. Running time of the different algorithms as function of s_{new} for uniform distribution.

From the above experiment results, we can see that the SD heuristic gives satisfactory results in terms of the number of kicked bundles, as compared with the SD optimal algorithm. At the same time, SD heuristic significantly outperforms the traditional techniques, e.g., best fit and worst fit. This naturally raises the question of whether SD heuristic is practical in terms of running time, as compared to the traditional techniques. To answer this question, we carried experiments that compare the execution time of the different algorithms. The results are shown in Figure 8. The y-axis shows the response time of each algorithm in milliseconds; the x-axis shows the number of services running in the gateway. As we see from this figure, while the optimal algorithm is significantly slower than the others, SD heuristics performs very well compared to the traditional techniques in terms of their running time. It is just what we have expected.

6. Conclusions

In this chapter, we have considered the problem of managing services and bundles in home gateways with limited amount of main memory. Because of the different architecture of home gateway using OSGi from the traditional computer architecture, a key difference between our problem and the traditional memory management is that the dependencies among different services have to be taken into consideration for a higher customers' satisfaction.

We use a dependency graph to model the relationship among services. This chapter proposes two algorithms. The first one is an extension of Knapsack problem which finds the optimal solution in a polynomial time. The second one is a heuristic that spans the dependency graph and tries to free the required amount of memory while minimizing the number of terminated services. We compared the proposed techniques with the traditional memory management algorithms such as the best fit and worst fit. Our experimental results indicate that SD (service dependency) heuristic is a good candidate for use in practical environments, as its performance is close to the optimal solution in terms of the number of stopped services. SD heuristic performs much better than the traditional memory management techniques. From the execution time point of view, SD heuristic is almost as fast as the traditional memory management techniques.

In this chapter, we have not taken into account of the priorities of different services. Our future work will focus on extending the proposed model to include the service priority. Different services may have different priority which determined by their specific characteristics or set by users. For example, an Internet game should not force out from the gateway a home security service (which is much more important than the internet game). Each service defines a priority value that reflects the importance of this service relative to other services. We will introduce the priority as a new factor in both the heuristic and the optimal solution.

7. References

Ali, M., Aref, W., Bose, R., Elmagarmid, A., Helal, A., Kamel, I., &Mokbel, M. (2005). NILE-PDT: A phenomenon detection and tracking framework for data stream management systems. In Proceedings of the Very Large Data Bases Conference, August.

Binstock, A. (2006). OSGi: Out of the gates. Dr. Dobb Portal, January.

Bottaro, A., Gérodolle, A., & Lalanda, P. (2007). Pervasive service composition in the home network. In Proceedings of the 21st International IEEE Conference on Advanced Information Networking and Applications, Niagara Falls, Canada, May.

Garey, M., & Johnson, D. (1979). Computers and intractability. New York: Freeman.

Helal, A., Mann, W., El-zabadani, H., King, J., Kaddoura, Y., & Jansen, E. (2005). Gator Tech Smart House: A programmable pervasive space. IEEE Computer, 38(3), 50–60.

Ishihara, T. (2006). Home Gateway architecture enabling secure appliance control service. In Proceedings of the 10th International Conference on Intelligence in Network (ICIN'06).

Ishihara, T., Sukegawa, K., & Shimada, H. (2006). Home Gateway enabling evolution of network services. Fujitsu Science Technical Journal, 24(4), 446–453.

Jain, K., & Vazirani, V. V. (2001). Approximation algorithms for metric facility location and k-Median problems using the primaldual schema and Lagrangian relaxation. Journal of the ACM, 48 (2), 274–296.

Jansen, E., Yang, H., King, J., Abdul Razak, B., & Helal, A. (2006). Acontext driven programming model for pervasive spaces. In 4thInternational Conference on Pervasive Computing, May.

Johnson, D. S., & Niemi, K. A. (1983). On Knapsacks, partitions, and a new dynamic programming technique for trees. Mathematics ofOperations Research, 8(1), 1–14.

King, J., Bose, R., Pickles, S., Helal, A., Vander Ploeg, S., & Russo, J.(2006). Atlas: A service-oriented sensor platform, the 4th ACMConference on Embedded Networked Sensor Systems (Sensys), Boulder, CO, USA.

Lee, C., Nordstedt, D., & Helal, A. (2003). OSGi for pervasive computing. the Standards, Tools and Best Practice Department, IEEE Pervasive Computing, A. Helal, Dept. Editor, Volume 2, Number 3, September.

Maples, D., & Kriends, P. (2001). The open services gateway initiative: An introductory overview. IEEE Communication Magazine, 39(12), 110–114.

Margherita2000, The first washing machine on the Internet. http://www.margherita2000.com/sito-uk/it/home.htm.

Microsoft Corporation, (2008). Universal plug and play device architecture reference specification, version 2.0. http://www.upnp.org/.

Ryu, I. (2006) Home network: Road to ubiquitous world. In Proceedings of the International Conference on Very LargeDatabases (VLDB).

Silberschatz, A., & Peterson, J. (1989). Operating system concepts. Boston, MA: Addison Wesley.

Sommers,F.(2006). Dynamic clustering with Jini Technology. www.artima.com/lejava/articles/dynamic_clustering.html, January.

Sun Microsystems Inc. (2007) Jini architectural overview. http://www.jini.org/.

The OSGi Alliance. (2009). The OSGi Service Platform release 4 core specification Ver 4.2. http://bundles.osgi.org/browse.php, September.

Vidal, I., García, J., Valera, F., Soto, I., & Azcorra, A. (2006). Adaptive quality of service management for next generation residential gateways. In Proceedings of the 9th International conference on Management of Multimedia and Mobile Networks and Services, Ireland, Dublin.

Watanabe, K., Ise, M., Onoye, T., Niwamoto, H., & Keshi, I. (2007). An energy-efficient architecture of wireless home network basedon MAC broadcast and transmission power control. IEEETransaction on Consumer Electronics, 53(1), 124–130.

Zigbee Alliance, (2004). Zigbee specification: Zigbee document 053474r06 Version 1.0, 14 Dec.

10

Virtual Place Framework for User-centered Smart Home Applications

Jumphon Lertlakkhanakul and Jinwon Choi
Yonsei University
Republic of Korea

1. Introduction

In smart home systems, building facilities and networked appliances communicate and operate with the others to perform the home services. Generally, these services are invisible and contain a series of diverse functions handled by separated devices. In fact, smart home can be regarded as a 'smarter' version of home automation system by adding a context-aware ability. Ma et al. (2005) defines 'smart space' as a space that must have some kinds of levels of abilities of perception, cognition, analysis, reasoning and anticipation about a user's existence and surroundings, on which it can accordingly take proper actions. In such an environment, computational intelligence can be regarded as being embedded into user's environment, including the space around the users (Weiser, 1991), rather than into the individual devices. Depending on the level of context adaptation, a smart home may fully controls the environment automatically or lets the occupants run services and manipulate the space on their own.

In architectural practice, it has been realized that there is a considerable gap in the communication between architects and users which always brings about the failure in real design or built environment in which users do not satisfy and never expect. Some serious cases found after early occupancy need to be solved through retrofitting which is a common and costly process we (architects) try to obviate (Palmon et al., 2006). Architects who come up with design solutions fail to deliver their ideas to users completely. The problem usually stems from a fact that users cannot imagine how the design will be emerged after construction phase. Unlike architects, users are not trained and their comprehension in three-dimensional space is limited. Consequently, such problems will become more considerable in case of smart home where a lot of interconnected equipments and complicated services are installed. These complex and invisible services can lead to the difficulty in occupants' role over the whole smart home life-cycle beginning from the design process to the occupancy stage. As any interactive home will be eventually used by end users, providing a method to enhance their participation and comprehension on how smart equipments and service will be installed as well as be operated will became major forthcoming issues in smart home industry. The efforts towards user-centered services can be found in a small number of projects such as Barkhuus and Day's study of user acceptation to context-aware service (Barkhuus and Dey, 2003) as well as Leijdekkers and

Gay's user profile service (Leijdekkers and Gay, 2005). Nonetheless, there is no research which applies the user-centered approach to architectural design stage so far.
The goal of this paper is to propose a new framework which allows smart home designers and smart home users to collaborate. The designers can configure spatial interaction caused by context-aware services and let the users to experience the home services during the design stage. This can be regarded as an interface which connects the occupants to the smart sensing environment. To do so, a new integrated framework between Context-aware Building Data Model, Virtual Reality (VR) and web service is introduced in this paper. The new building data model is created base on Structured Floor Plan (Choi et al., 2007) to handle the interactivity and the complexity of smart home services. VR is applied to visualize invisible and pervasive sensing networks running in the background as well as providing an immersive environment for spatial interaction manipulation. Lastly, the web service technology is utilized to increase the system accessibility and to imply inter-connectivity to smart home equipments. Therefore, this paper examines how to create and to implement virtual space using VR technique as a platform to simulate smart home service configuration.
In this paper, we propose a series of smart home platforms which enables home users to experience smart virtual place through the Internet. In particular, our interactive virtual place is different from conventional 3D space in that the created virtual place embodies spatial context-aware information including spatial relationship among entities, activities and users. Avatars controlled by users can explore and perform a set of related activities according to the current context resulting in the change and the interaction of virtual place. Consequently, the system can be used to simulate not only how space will look like but also how users interact with the smart environment based on predefined scenarios.
To achieve our goal, our research is conducted through following processes. First, similar and related systems are analyzed to indicate the research direction and the evaluation model. Second, essential elements to construct the virtual smart environment are extracted. Third, a novel place data model is constructed. After that, a series of smart home prototypes composed of 'PlaceMaker', 'V-PlaceLab' and 'V-PlaceSims' are developed based on the place data model. At the end, the overall processes to demonstrate how smart home designers and users can utilize the prototypes are discussed.

2. Related Works

To propose a new smart home framework, it is necessary to comprehend various related subjects including smart home environment, VR and behavioral research. This section describes state-of-the-art technology related and clarifies our research position developed with a different approach.

2.1 Smart Home Environment

According to Chen and Kotz (2000), context-aware services can be classified as passive or active. Active context-aware services are those that change their content autonomously on the basis of sensor data whereas passive context-aware services only present the updated context to the users and let them specify how the application should change. Likewise, smart home can also be categorized as passive smart home and active smart home depending on the services provided.

Passive smart homes which react to occupancy command are widespread whereas active smart homes, those demand interaction and invite guidance have not been vastly adopted in the housing market yet. Examples of active smart home are The Aware Home (Kidd et al., 1999), Gator Tech Smart Home (Hetal et al., 2005), Toyota Dream House PAPI (Sakamura, 2005) and NICT's Ubiquitous Home (Minoh and Yamazaki, 2006). Accordingly, most active smart homes are found in R&D projects as it requires greater advanced and costly technology that cannot be commercialized at the moment. Nonetheless, the barrier of smart home application does not stem from only the cost problem. Indeed, the pervasiveness and the invisibility of devices and their working capacity also come with trade-offs.

For active smart home, The Aware Home (Kidd et al., 1999) is one of the first-generation laboratory houses for elderly developed at Georgia Institute of Technology. The research home was simultaneously inhabited by elderly people as well as tested and monitored by researchers. The research goal was to apply ubiquitous computing for everyday activities. Another similar project is Gator Tech Smart House (Hetal et al., 2005) developed by Mobile and Pervasive Computing Laboratory at University of Florida. With extensible technology based on OSGi framework, the goal of this context-aware home was to create an 'off-the-shelf' smart house which the average user can buy, install, and monitor without the aid of engineers. Compared with The Aware Home, Gator Tech Smart House is more appliance-oriented. Various smart functions for smart home appliances, home security system and home assistant service have been being developed. In Japan, the same movement in context aware home has been well recognized at Toyota Dream House Papi (Sakamura, 2005). The home has been developed under 'TRON' project, a long-term project since 1984 aimed at creating ideal computer architecture (http://tronweb.super-nova.co.jp). The main goals for the smart home were to design and to realize an environmentally friendly, energy saving intelligent house design in which the latest ubiquitous network computing technologies created by the 'T-Engine' project (Sakamura, 2006) could be tested and further developed. Another recent example of active smart home in Japan is Ubiquitous Home (Minoh and Yamazaki, 2006) developed at National Institute of Information and Communications Technology (NICT). Similar to The Aware Home and Gator Tech Smart House, families were invited to stay and test home services in the living laboratory. However, the home was applied with 'Mother-Child' metaphor having robots to take care of occupants. Unconscious type home robot controlled all services in the background where as visual type interface robots were used to communicate with the occupants.

Regardless of the different scopes and applications, common characteristics of above active smart homes have been noticed as follows; (1) Building components and networked appliances communicate and operate with the others to perform context-aware services. (2) Generally, smart services are invisible and contain a series of diverse functions handled by separate devices. (3) The home is capable of identifying and predicting its occupants' actions by means of sensors and actuators then commit actions on behalf of them by means of Artificial Intelligence (AI). Considering these smart home cases, it is obvious that current research and development on smart home aims at creating the home capable of understanding its inhabitant as much as possible. However, this research argues that an opposite approach is more important and must be taken into account.

In addition, there are no current smart homes which can solely control the environment so far. Some smart home systems like NICT's Ubiquitous Home (Minoh and Yamazaki, 2006) and LG HomNet (http://www.lghomnet.com) apply the concept of 'Home Mode' to operates all smart services according to the current mode. For example, a home may offer sleep mode, wake up mode, away mode, etc. In fact, the operation for each mode may vary from one user to the others. In other words, each user may have individual preferences on how the smart home should operate or be operated. Therefore, instead of letting the home understand the inhabitants, it is more important to acknowledge users on how the smart home can work and be operated at the moment.

2.2 Virtual Reality in Simulation

According to Weiss and Jessel (1998), one of the cardinal features of VR is the provision for a sense of actual presence in, and control over, the simulated environment. Simulation of spatial reality has a key role in order to duplicate the experience of real space (Oxman et al., 2004). VR platforms, therefore, have been extensively developed and exploited for simulating real space using virtual environment. In particular, under certain conditions such as occur when a task is more meaningful, interesting or competitive to the user, the level of presence is generally improved, even in the absence of high immersion (Nash et al., 2000). Moreover, Oxman and colleagues (2004) introduced three design paradigms to induce presence in virtual environment: task-based design, scenario-based design and performance-based design. In fact, such paradigms can be found in situation simulation games such as 'The Sims2' (Ma et al., 2005) in which each user performs ordinary tasks imitating the life in real world. The game playing depends on emotional and behavioral characteristics of multiple users through complex scenarios. Oxman's paradigm, therefore, can explain why the level of presence in a situation simulation game is high enough to enable game players to immerge and to enjoy the interaction in virtual environment. Apart from these studies, a number of outstanding VR simulation platforms have been developed revealing the same tendency. FreeWalk/Q (Nakanishi and Ishida, 2004) developed at Kyoto University was a platform for supporting and simulating social interaction in Digital Kyoto City. Its goal was to integrate of diverse technologies related to virtual social interaction, e.g. virtual environments, visual simulations and lifelike characters (Prendinger and Ishizuka, 2004). In FreeWalk/Q, lifelike characters (referred to both avatar and agent) enable virtual collaborative event such as virtual meeting, virtual training, and virtual shopping in distributed virtual environments. Furthermore, the system utilized 'Q', an extension of a Lisp programming language called 'Scheme' as a scenario description language for describing interaction scenarios between avatars and agents. Unlike the research mentioned above which emphasizes user-user interaction or user-agent interaction, our approach focuses on the interaction between user and virtual space to enable context-aware services and functions as found in physical smart space.

2.3 Virtual Reality in Behavioral and Architectural Simulation

Meanwhile, there have been the attempts to study about human behavior in a certain kind of place using VR. Wei and Kalay (2005) developed a behavioral simulation platform embedded with usability-based building model. Their original building model created in DXF format is converted into scalable vector graphics (SVG) format then appended with non-graphical information. Such model enables virtual users as agents to perform specific behaviors autonomously for each spatial building entity. Our research also applies similar concept to this spatial building model. It is, nevertheless, developed upon Spatial Context-aware Building Data Model (Lertlakkhanakul et al., 2006). Another research by Palmon and colleagues (2006) introduced how a specific group of users such as people with disabilities can apply VR technology for a pre-occupancy evaluation. This project involved in the design of home environment before the construction phase. The system utilized an interaction with virtual environment verifying the ease of navigation and object usability using a joystick. However, the interaction level between space and users through their avatars was rather limited to collision detection and change in object attributes. Our research goal is also to create a spatial interaction management tool focusing on smart home environment. Hence, it requires concentrating on a higher level of human-space interaction in virtual environment.

2.4 Virtual Reality for Smart Environment

Recently, a new concept to combine two distinct paradigms called 'Ubiquitous Virtual Reality' (U-VR) has been introduced. According to Kim et al. (2006), VR focuses on the activities of a user in a Virtual Environment (VE) that is completely separated from a Real Environment (RE). On the other hand, Ubiquitous Computing (ubiComp) focuses on the activities of a user in a RE. Although VR and ubiComp reside in different realms, they have the same purpose, i.e. to maximize the human ability. Pfeiffer and partners (2005) presented a new method for remote access of virtual environments based on established video conferencing standards. A wide range of clients, from mobile devices to laptops or workstations, were supported enabling the virtual environments ubiquitously accessible. In addition, Kim and his colleagues (Kim et al., 2006) described and explored U-VR in a broader sense related to ubiComp. By supplementing the weaknesses of VR with the help of ubiComp, they looked for ways to evolve VR in ubiComp environments and purposed a demonstrated platform called Collaborative Wearable Mediated Attentive Reality. Nevertheless, our research is different from their research in that, the concept of U-VR is not applied to the interoperability in communication method and collaboration. Rather, it investigates how we can increase the usability of smart home context by means of VR.

3. The Building Data Model for Smart Home

In this paper, we explore how to create and to implement virtual space using VR technique as a platform to simulate smart home configuration. Due to the advancement of technology installed, smart homes require a novel simulation tool to help users realize designed smart home configuration before construction phase. Unfortunately, traditional CAD models possess only graphical/geometric information of design element (Wei and Kalay, 2005). They are lack of spatial information and other non-geometric information needed in order to create the smart virtual environment which can interact with virtual users.

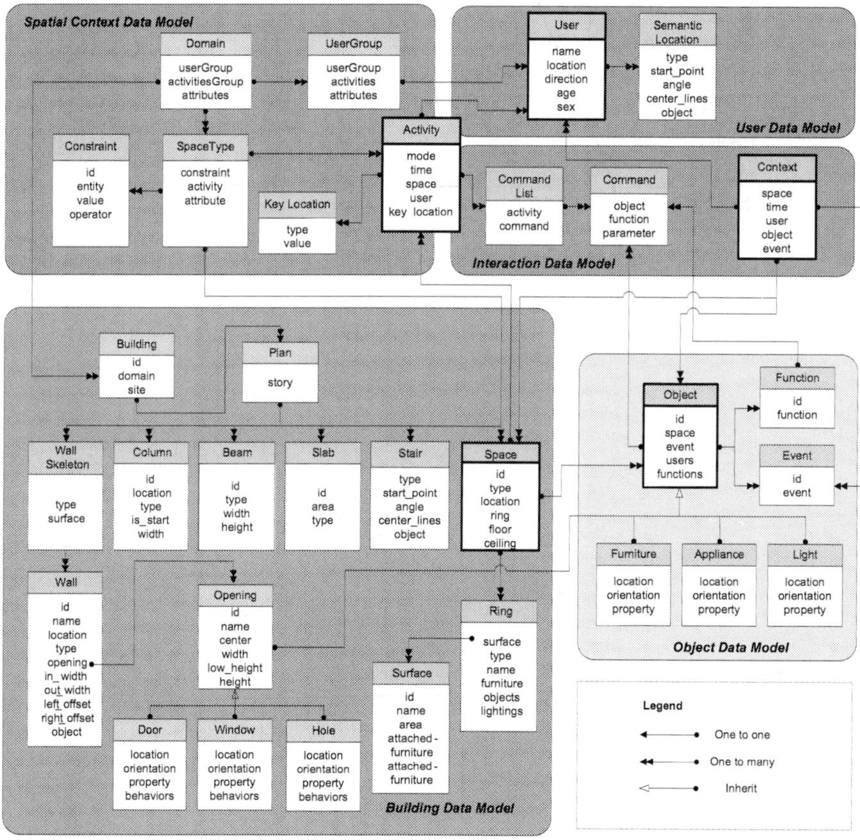

Fig. 1. Spatial Context-aware Place Data Model

At this stage, a method to initiate smart characteristic of virtual environment is needed to be investigated. According to our goal, human-space interaction plays a key role in the simulation. Besides, Ma (2005) defines 'smart space' as a space that must have some kinds of levels of abilities of perception, cognition, analysis, reasoning and anticipation about a user's existence and surroundings, on which it can accordingly take proper actions. To be concise, the most basic characteristic of smartness is the context semantics and awareness for a real space or environment (Dey, 2001). Based on such definition, a novel data model called Spatial Context-aware Place Data Model has been defined. Unlike traditional CAD models, it embodies both geometry information and semantic information including spatial relationship among humans, objects and spaces. In particular, it enables the concept of virtual place. Places differ from mere spaces in that they embody social and cultural values in addition to spatial configurations. It is the concept of place, not space that connects architecture to its context and makes it responsive to given needs (Kalay, 2004). Figure 1 illustrates the structure of spatial context-aware place modeling. Its main components are described in this section.

3.1 Virtual Building Data Model

Built environment consists of a large number of design elements, such as walls, doors, windows, stairs, columns, etc. A virtual user needs to perceive and understand them to behave properly (Wei and Kalay, 2005). Furthermore, humans do not perceive architectural space as an image, but as a hierarchical composition of various elements (Lee et al., 2004). Therefore, our virtual building data model includes spatial information to explain the configuration and hierarchy of spatial components based on the concept of Structured Floor Plan (Choi et al., 2007). Spatial reasoning is the main advantageous feature of this model. It systematically enables virtual user to perceive and to recognize the space using hierarchical relationship and spatial connectivity among building components. From top-down approach, the member classes of the virtual building modeling range from 'Building' to 'Surface'. Among them, 'Space' functions as a main interface class connected to other classes.

3.2 Virtual Object Data Model

Unlike traditional CAD models, our model does not only include building components but also objects including furniture, equipments and appliances. In addition, our model is capable of spatial context-awareness. Thus, smart objects also contain their own functions and status to interact with users and other objects. Such interaction is activated according to specific events performed by the virtual users in the same manner as occurred in the real world by means of sensors installed in smart objects. Each object must belong to at least one space (except doors and windows), enabling them to communicate with other entities.

3.3 Spatial Context Data Model

The modeling of spatial context handles additional non- geometric information attached to a space. It describes typical characteristics and spatial configuration for the built environment. 'Domain' stores spatial information of building type in the same manner as 'Space Type' does for space. Generally, domain and space type for each space are unique. For example, library and home (different domains) are so different in spatial configuration. They require disparate spaces, activities, area used by different types of user. Living room and dining room (different space types) need different furniture, temperature and location regardless of the same 'home' domain. In short, our spatial data model performs as a typical spatial knowledge base of any agent based system.

3.4 Virtual User Data Model

For the sake of offering appropriate services to each user in ubiquitous computing space, the system must be capable of storing and retrieving personal information precisely since each user may have unique characteristics and behaviors. Even a user can have dissimilar preferences under different situations. The personal preferences and needs, persons to interact with, and sets of devices to control by each individual, define one's personal communication space (Arbanowski et al., 2001). Such personal information is stored in the Virtual User Data Model and handled by 'User' and 'Activity' classes.

3.5 Interaction Data Model

All potential interaction channels between a virtual user and the smart virtual environment are taken into account. Interaction in the virtual environment could take place by means of Interaction Data Model. It functions as an interface between the Virtual User Data Model and the others. In other words, it motivates the concept of human-centered service by applying context-aware ability. For example, a virtual user can perform specific actions with each object by the connection between 'Event' class stored in the Virtual Object Data Model and the User class. This enables a user to sit or lay on a bed for example. This linkage is created and handled by 'Context' class. Context is any information that can be used to characterize the situation of an entity (Ma et al., 2005). It serves as the key transaction and the initial status for any possible interactions by connecting all the components such as space, user, object, activity and event. Once a specific event performed by a user is detected, all related activities will be retrieved as the user's potential goals. Each activity contains a set of commands for operating all related objects and services. More details on how interaction model works are explained at the end of this paper.

4. Virtual Smart Home Framework

In the previous section, the building data model for virtual smart home is introduced. It serves as the kernel to enable context-aware interaction. The system is capable of specifying who is doing what action at which area with which object for which purpose. In addition, the spatial network can be used to create location based service as well as enabling spatial reasoning. However, in order to create a virtual smart home environment for simulation, it requires developing the virtual environment itself with two uppers layers over the data model. Figure 2 shows the holistic framework of the virtual smart home. At the top level, the web layer connects the users to the virtual smart home model. In the intermediate level, agent layer interacts (reasoning) with user actions in the virtual environment and has access to the data model located at the bottom layer. The agent layer can be regarded as invisible robots or a smart home server in physical smart home cases. The process to create the virtual environment and the description of the two layers are discussed in this section.

4.1 Virtual Place Modeling Process

On the lowest layer, the overall mechanisms of the virtual architecture are motivated by means of the Building Data Model for smart home. Spatial Context-aware Data Model is capable of storing semantic information and spatial context for smart architecture apart from geometric data. The virtual place making process begins with using PlaceMaker, our spatial context-aware CAD modeling system, to design a virtual home by an architect according to the users' preferences. Therefore, the output model contains both geometric and semantic spatial information including user and activity lists. The next step is to insert smart objects including their functions and events using a tool so called 'V-PlaceLab'. Here, avatars are inserted as simulated users to create and test scenarios of spatial usability. The linkage between user activity and object operation is also applied at this stage. Finally, the smart home model is exported and embedded in a web page then uploaded into an online platform so called 'V-PlaceSims'. More details on PlaceMaker, V-PlaceLab, and V-PlaceSims can be found in Section 5. Once a user has registered and input his/her personal information, the entire virtual model is ready to be utilized. Figure 3 shows the virtual place making process.

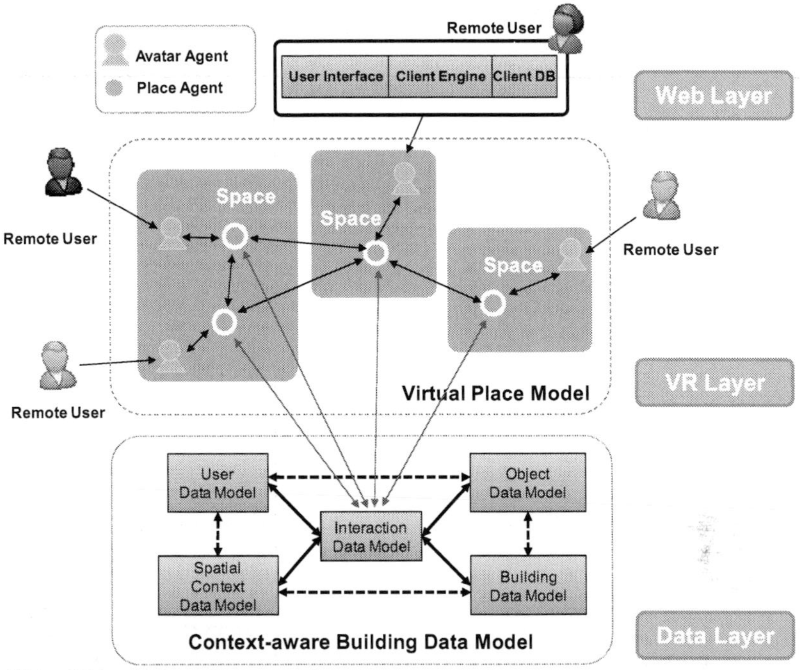

Fig. 2. Virtual Smart Home Framework

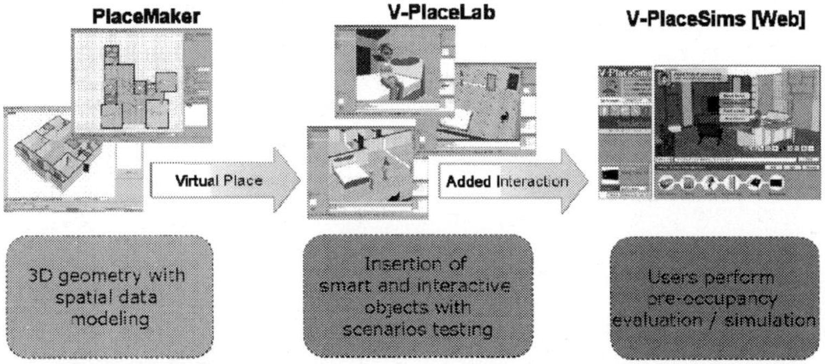

Fig. 3. Virtual Place Making Process

4.2 Home and User Agents

A multi-user environment is created using client-server technology. The system model comprises three modules; client module, web application and server. A user interacts with other users and the virtual environment through user interface layer provided in a client browser. Avatar agent and place agent running in the web application layer sense the ad-hoc context and commit changes to the virtual world.

Our main computation processes for the agents are created based on UCVA Agent Model (Maher and Gu, 2002). Agents are defined as 'Reactive' agents reasoned response to an expected event in the virtual world. Their main processes are sensation, perception and action. Place agent takes care of context-aware reasoning including social event definition as well as spatial interaction control between users and the virtual space. It collects all context information for each space such as space type, number of user, current activity compared with social event definition to define current event for a single space. In contrast, avatar agent keeps handling user interaction. Each user has his or her own avatar agent. It senses the user's action then transmits the data to the place agent and waits for the response.

4.3 Web Layer and Graphics User Interface

At the top level, web service with server-client topology is constructed to enable the online simulation. Users connect to the web server using HTTP protocol through the system homepage. On one hand, the dynamic WebPages are written in ASP script embedding ActiveX control (stage control in the interface) implemented using C++ language. The control engine can be automatically downloaded and installed on the first visit. The engine task is to handle user interaction and to render the virtual environment on the ActiveX control. The client side also includes texture library, avatar expression and motion database used for rendering purpose. On the other hands, the server locates the main computation engine where place and avatar agents are registered to update the virtual place context and gain connection to the database on the lower level. The ActiveX control works with other dynamic controls on the web pages to enable user actions and to display the updated virtual smart home environment using stream socket to synchronize clients with the server.

5. Virtual Smart Home Applications

Based on the virtual smart home framework described previously, we created a series of virtual smart home applications to realize the framework. The series includes PlaceMaker, V-PlaceLab and V-PlaceSims. Descriptions of each application are explained in this section.

5.1 Place Maker

PlaceMaker is our modeling tool in which intelligent building model bound with the spatial context can be created. While creating an instance of building components in PlaceMaker, spatial context is automatically generated from library along with the geometric data. The system is designed to be capable of elaborating various kinds of building domain and space type. Buildings created within different domain contain different spatial characteristic and spatial context. Figure 4 shows the entire interface of PlaceMaker. There are three modes of visualization; Two-dimensional mode (2D mode), Three-dimensional mode (3D mode), and Print mode. Users can freely switch the visualization mode back and forth among all modes. By default, the 2D mode is set as the active mode. In this mode, users can create a space by drawing enclosed walls. A space can be composed of building components such as wall, opening, furniture as described in the Spatial Context-aware Place Data Model. All instances can be modified through their parameters in real-time manner with dialogs. Basic CAAD modeling and operation tools are also provided. Space type must be assigned to bind them with spatial context. By doing so, this enables various features including constraint-based design, automation in spatial network and procedural modeling. More information on PlaceMaker can be found in Lertlakkhanakul and colleagues (2006).

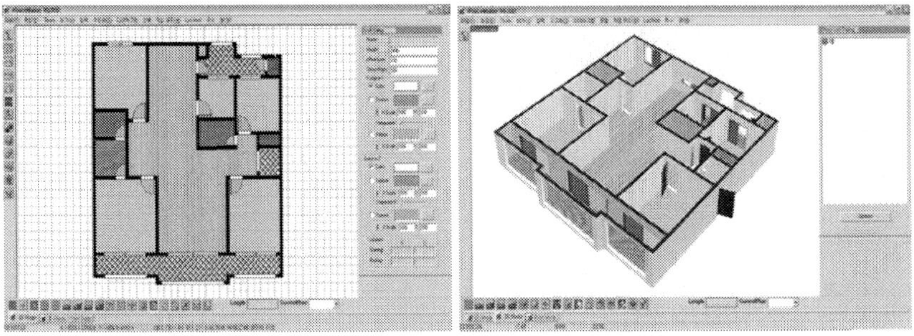

Fig. 4. The graphical user interface of PlaceMaker: 2D mode (left) and 3D mode (right)

5.2 V-PlaceLab

Not only spatial information, user information is also the main part to construct the virtual place embedding human-space interaction. Based on the Place Data Model, the next task is to include activities and their location in our model. Activity entity combines all information from user, space and location to form a place. In other words, it is possible to retrieve information about available activities, their locations and users which are useful for any context-ware or location based systems. To do so, V-Placelab (Lee et al., 2008) has been developed for this purpose. It is a simulation tool for smart home which enables designers to create scenario demonstrating how users can utilize the space. Scenario is created by setting the interaction among spaces, objects and human behaviors. Smart objects can be inserted into the model previously created by PlaceMaker. Each object not only contains geometry but also information of activity and its key location. For instance, sofa functions as an action point for watching TV and reading activities while bed is a location for sleeping. All information are merged and united into one virtual building. Each activity is associated to a home service and commands in the Interaction Data Model. With this interaction embedded in the virtual model, smart home designers can test the home functions by means of avatar control. Invisible services are represented with balloons over avatars. Figure 5 shows the screen shots of V-PlaceLab. Series of avatar actions can be recorded as scenarios and played-back as animation. The end result of virtual smart home is uploaded to V-PlaceSims Server for the collaboration between designers and smart home users.

Fig. 5. The graphical user interface of V-PlaceLab

5.3 V-PlaceSims

V-PlaceSims is developed as an online virtual environment platform for design collaboration between the project architect and the users. Our research emphasizes the collaboration process of smart home design. At first, a well-designed smart virtual home equipped with smart objects is provided by the architect. ActiveX technology embedded in Active Server Pages (ASP) format has been chosen to be the development platform as mentioned previously. After uploading the complete virtual model into our web server, users can login and control their avatars to explore the virtual smart home through their Internet connection. The users can navigate and interact with the virtual environment as thoroughly explained in the following section. The graphical user interface of V-PlaceSims can be divided into four main parts; Stage, Property panel, Command panel, and Action panel. Occupied the largest displayed area, 'Stage' illustrates the interactive virtual environment. It also includes a chatting textbox at the bottom and a dialog on the top for a communication. Located on the left area, 'Property' panel includes Status panel, Mode-setting panel, and Object Properties panel. A user can perform various functions including setting their information, preferences and preferred activities for each room as well as running a simulation. In addition, architects are privileged to modify the spatial configuration such as inserting, removing and moving objects. Located at the screen bottom, 'Command' panel displays how the virtual model interacts with users by showing the sequence of home networking commands. Lastly, 'Action' panel is located at the bottom showing available actions provided for each mode. Figure 6 illustrates a screenshot of V-PlaceSims.

Within V-PlaceSims, multiple users can explore the space simultaneously as well as exchange their opinions through designated chatting textbox and utilize their (avatar's) body expressions. Likewise, users can also receive assistance and send their feedback to the designer using the same communication channels. Moreover, users can interact with the smart environment and also possess a certain permission level to manage smart service configuration. Figure 7 shows the holistic view of collaboration model in V-PlaceSims.

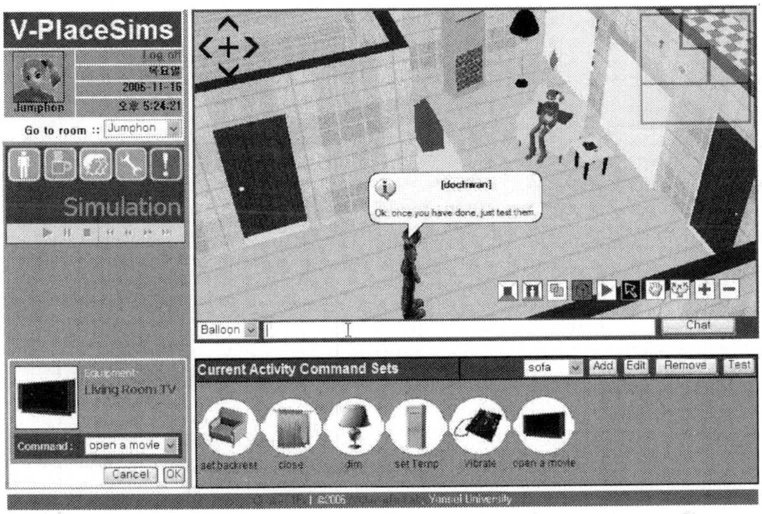

Fig. 6. A screenshot of V-PlaceSims

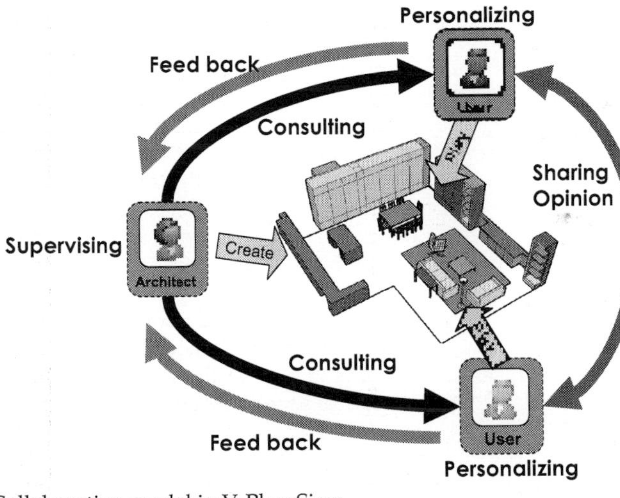

Fig. 7. Collaboration model in V-PlaceSims

6. Demonstration

V-PlaceSims provides a set of functions divided into five modes. They are Avatar, Activity, Security, Space-editing, and Simulation modes. Users can set the current mode using navigation controls located in the Mode-setting panel. Details of each mode are summarized as follows; (1) At 'Avatar' mode, users can choose their avatar, edit their preferences and profiles such as favorite TV programs, water temperature for shower, favorite song for morning call, room temperature for sleeping, and so on. (2) At 'Activity' mode, users can specify what type of activity they would like to do for each room. The activities are grouped according to the home mode; such as sleeping, cooking, and going-out modes. (3) At 'Security' mode, users can modify their account information such as username and password. The system administrator can also edit the user permission in this mode. (4) At 'Space-editing' mode, users are capable of adding, transforming (move and rotate) or deleting furniture as well as setting room textures. Thus, the furniture layout can be freely design by the users. (5) At 'Simulation' mode, users can change to this mode after completing the setting mode. The purpose of this mode is to let the designer or the users customize how the home components and appliances can be used according to the users' needs as each user may have personal preferences in living style. This can enable, for example, how services in a bedroom including illumination system, air-conditioning system and audio system can function together to support a user when going to bed for sleeping mode (e.g. as the user lays down on the bed at night). Here, the designer can acknowledge the users on how the smart objects can be operated or let them decide how they want to user the space on their own.

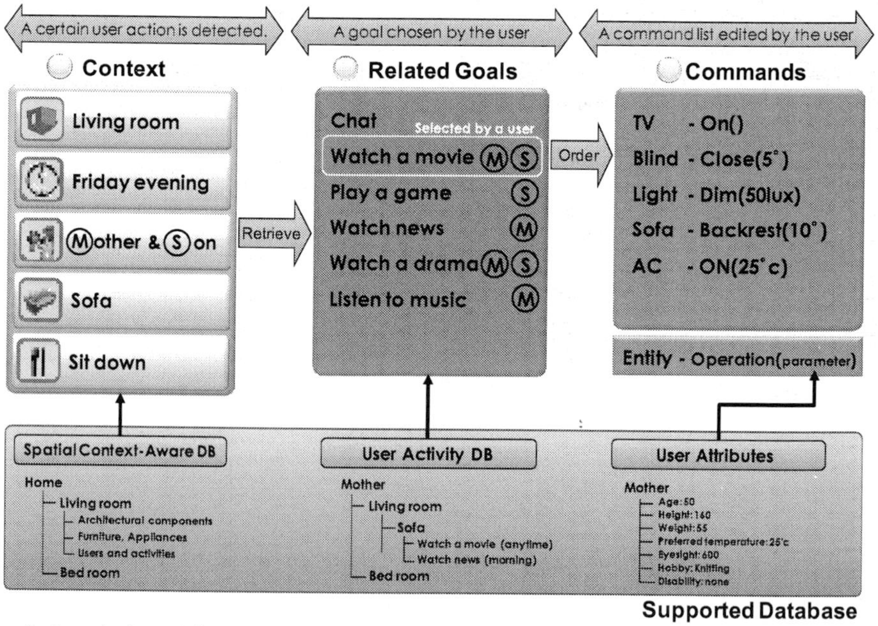

Fig. 8. A typical spatial interaction scenario

Following content demonstrates how V-PlaceSims is utilized by users through a scenario as shown in Figure 8. A mother and her child enter the virtual living room (through their avatars) on Friday evening. As a smart home designer, the architect guides the users to sit on a sofa located in that room. As soon as they sit on the sofa, the sitting event activates smart interaction processes. This procedure has been set by the architect in advance. Then the system searches for all related activities from users' personal information. All of them are displayed on the stage right after they have sat on the sofa. Note that, only the main user set by the architect has right to select his/her preferred activity when there are many users activate the same event at a single time. Then, the mother as the main user selects an activity as her goal. After that, the system lists up all designated object commands according to the chosen goals. Each object command contains both functions and parameters. The parameters are also automatically defined based on the user's personal information such as preferred room temperature, lighting illumination and color, music volume, and so on. Once the mother has seen a command list, she can also rearrange the order as well as remove and add new command in the panel. After fixing all commands, she can run the simulation and see the result changing in the virtual environment setting. Example screenshots of the simulation are shown in Figure 9. Currently, the replacement of furniture model (e.g. to replace an opened curtain with a closed one), the use of geometry representation of smart service (spheres and lines over objects) and balloon message are being used to response with users and shown in Figure 9 and 10.

A B C

Fig. 9. Screenshots of V-PlaceSims: (A) Space Edit Mode, (B) and (C) Simulation Mode

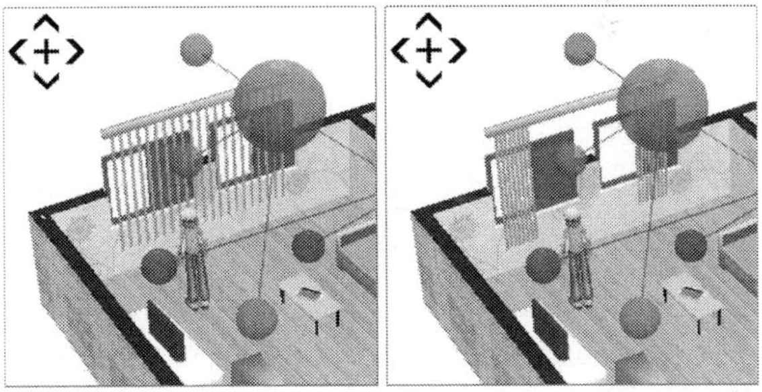

Fig. 10. Visualization of smart home services by replacing objects and using network of spheres

7. Conclusion

In this paper, a new virtual place framework has been proposed to enhance the collaboration between smart home designers and end users. The framework incorporates a range of processes beginning at defining a generic data model following by the development of virtual place applications including PlaceMaker, V-PlaceLab and V-PlaceSims. New interactive characteristics of building data model have been introduced. In designers' point of view, our framework offers a new method to create virtual smart home model embedded with human-space interaction and service information. Building models created by PlaceMaker can be placed with smart objects holding activity information. With V-PlaceSims, a smart home user can control his/her avatar to interact with smart environment in such a manner that players enjoy 3D computer games. The interaction refers to a sequence of smart object commands evoked by certain user actions. On the contrary to agent-based simulation (no ad-hoc user interaction), our approach uses data input by actual users as avatars' actions and behaviors in virtual place. This leads to the paradigm shift in user role from a passive listener to an active actor. In addition, group collaboration is also possible in which users can give their feedbacks and update the spatial interaction instantly.

By means of the Context-aware Building Data Model, human-space interaction which is vital for simulating smart home functions and services is realized in the virtual environment. Together with VR, the platform is capable of visualizing invisible services, performing real-time interactions with the home and acknowledging the users how it can be configured and operated according to their individual needs. In addition, the web-based service increases the system accessibility and usability as users can log-in from anywhere to collaborate with each others. Because smart home is extremely difficult for users to gain insight of smart service configuration, our research position is to express a user-oriented approach by applying the concept of U-VR to smart home simulation model. The long term goal is to avoid considerable gap between the architect and the user in smart home design process. After all, our framework can deliver users the better comprehension in how smart and interactive virtual architectural models designed by architects will be constructed and utilized. Eventually, it is expected that the research will lead to the decline in design failures and problems in built environment during building occupancy period.

8. References

Arbanowski, S., Meer, S.V.D., Steglich, S. & Zeletin, R.P. (2001). The Human Communication Space: Towards I-centric Communications. *Personal and Ubiquitous Computing,* Vol. 5, No. 2001, pp. 34-37.

Barkhuus, L. & Dey, A. (2003). 'Is context-aware computing taking control away from the user? Three Levels of Interactivity Examined'. In *Ubicomp 2003.* pp. 149-156.

Chen, G. & Kotz, D. (2000). A Survey of Context-Aware Mobile Computing Research. Department of Computer Science, Darthmouth College.

Choi, J.W., Kwon, D.Y., Hwang, J.E. & Lertlakkhanakul, J. (2007). Real-time Management of Spatial Information of Design: A Space-based Floor Plan Representation of Buildings. *Automation in Construction,* Vol. 16, No. 2007, pp. 449-459.

Dey, A.K. (2001). Understanding and Using Context. *Personal and Ubiquitous Computing,* Vol. 5, No. 2001, pp. 4-7.

Hetal, A., Mann, W., Elzabadani, H., King, J., Kaddourah, Y. & Jansen, E. (2005). Gator Tech Smart House: A Programmable Pervasive Space. *IEEE Computer Magazine,* Vol. March 2005, pp. 64-74.

Kalay, Y.E. (2004). *Architecture's New Media,* Cambridge: The MIT Press.

Kidd, C.D., Orr, R., Abowd, G.D., Atkeson, C.G., Essa, I.A., MacIntyre, B., Mynatt, E.D., Starner, T. & Newstetter, W. (1999). The Aware Home: A Living Laboratory for Ubiquitous Computing Research. In *Lecture Notes in Computer Science.* Berlin/Heidelberg: Springer-Verlag. pp. 191-198.

Kim, S.W., Suh, Y.J., Lee, Y.H. & Woo, W.T. (2006). Toward Ubiquitous VR: When VR Meets ubiComp. In *International Symposium on Ubiquitous VR.* Yanji. pp. 1-4.

Lee, Y.G., Lee, I.J. & Choi, J.W. (2004). Location Modeling for Ubiquitous Computing Based on Spatial Information Management Technology. In *CAADRIA 2004.* Seoul. pp. 787-801.

Lee, Y.G., Lertlakkhanakul, J., Choi, J.W. & Kalay, Y.E. (2008). Dynamic Architectural Visualization based on User-Centered Semantic Interoperability. In *ACADIA 2008.* Minneapolis. pp. 406-415.

Leijdekkers, P. & Gay, V. (2005). Personalized Service and Network Adaption for Smart Devices. In *IEE Asia Pacific Conference on Communications 2005.* pp. 867-871.

Lertlakkhanakul, J., Do, S.R. & Choi, J.W. (2006). Developing a Spatial Context-Aware Building Model and System to Contruct a Virtual Place. In *DDSS 2006.* Eindhoven. pp. 343-358.

Ma, J., Yang, L.T., Apduhan, B.O., Huang, R., Barolli, L. & Takizawa, M. (2005). Towards a Smart World and Ubiquitous Intelligence: A Walkthrough from Smart Things to Smart Hyperspaces and UbicKids. *International Journal of Pervasive Computing and Communication,* Vol. 1, No. 2005, pp. 53-68.

Maher, M.L. & Gu, N. (2002). Design Agents in Virutal Worlds - A User-Centred Virtual Architecture Agent. In *Agents in Design.* Gero, J.S. & Brazier, F.M.T. (Eds.). MIT. pp. 23-38.

Minoh, M. & Yamazaki, T. (2006). Daily Life Support Experiment at Ubiquitous Computing Home. In *IPMU 2006.* Paris. pp. 534-540.

Nakanishi, H. & Ishida, T. (2004). FreeWalk/Q: Social Interaction Platform in Virtual Space. In *VRST 2004.* ACM. pp. 97-104.

Nash, E.B., Edwards, G.W., Thompson, J.A. & Barfield, W. (2000). A Review of Presence and Performance in Virtual Environments. *International Journal of Human-Computer Interaction 12,* pp. 1-41.

Oxman, R., Palmon, O., Shahar, M. & Weiss, P.L. (2004). Beyond the Reality Syndrome: Designing Presented in Virtual Environments. In *ECAADE 2004.* Orbak, H. (Ed.). Copenhagen. pp. 15-18.

Palmon, O., Sahar, M., Wiess, L.P. & Oxman, R. (2006). Virtual Environments for the Evaluation of Human Performance: Towards Virtual Occupancy Evaluation in Designed Environment. In *CAADRIA 2006.* pp. 521-528.

Pfeiffer, T., Weber, M. & Jung, B. (2005). Ubiquitous Virtual Reality: Accessing Shared Virtual Environments through Video Conferencing Technology. In *EG UK Theory and Practice of Computer Graphics 2005.* The Eurographics Association. pp. 209-216.

Prendinger, H. & Ishizuka, M. (2004). *Life-Like Characters: Tools, Affective Functions, and Applications*: Springer Verlag.

Sakamura, K. (2005). Intelligent House in the Age of Ubiquitious Computing, In A House of Sustainability: PAPI. In *A+U.* A+U Publishing. pp. 56-65.

Sakamura, K. (2006). Challenges in The Age of Ubiquitous Computing: A Case Study of T-Engine, An Open Development Platform for Embedded Systems. In *ICSE 2006.* Shanghai. pp. 713-720.

Wei, Y. & Kalay, Y. (2005). Simulating Human Behaviour in Built Environments. In *CAAD Futures 2005.* Vienna. pp. 301-310.

Weiser, M. (1991). The Computer for the Twenty-first Century. *Scientific American,* Vol. 265, No. 3, pp. 94-104.

Weiss, P. & Jessel, A. (1998). Applications of Virtual Reality to Work. *Work,* Vol. 11, No. 1998, pp. 277-293.